Computing Methods
in
Quantum Organic Chemistry

Computing Methods
in
Quantum Organic Chemistry

H. H. GREENWOOD
Computer Centre, University of Keele

WILEY — INTERSCIENCE
A division of John Wiley & Sons Ltd
London — New York — Sydney — Toronto

Library of Congress catalog card number 78–149572

ISBN 0 471 32640 2

Made in Great Britain at the Pitman Press, Bath

Preface

In spite of the rapid advances made in the development of molecular orbital theory, the method, even in its simplest form as applied to π-electron systems of conjugated molecules, has yet to gain acceptance as a universal language for the description and interpretation of molecular properties. Classical electronic theories and the method of resonance structures are still more widely preferred and used in practice. Yet molecular orbital theory is conceptually easier to understand than the resonance method, and rests on firmer theoretical foundations; it provides quantitative interpretations which are broadly equivalent to those given by the classical electronic theories, and offers a more complete description of the theoretical implications. The most serious drawback to the advancement of molecular orbital theory has, hitherto, been the volume of computational work involved in practical applications. The time taken to calculate, on hand machines, π-electron distributions for molecules of practical interest, even in the simplest molecular orbital method, has generally been prohibitive for all, except theoretical chemists, especially when qualitative methods of interpretation prove attractive and versatile in practice. The situation is now changing, as more computing power becomes available to research workers, and to undergraduates for teaching purposes, and it is within the context of a growing interest in the use of computers that this book appears.

The book is intended to provide a practical guide to the use of computers for solving problems described in terms of π-electron molecular orbital theory. It covers the Hückel method, the self-consistent field method as applied in the approximation introduced by Parr, Pariser and Pople, and a configuration interaction method for calculating singlet and triplet excited states. It is not primarily concerned with the quantum mechanical foundations of molecular orbital methods, nor with the derivation of the equations; neither does it attempt to give an account of the success of MO methods in predicting and describing a variety of molecular phenomena. These are matters which have already been considered extensively elsewhere in existing textbooks and journals. Instead it attempts to show how computers may be used to study both the models which are introduced in molecular orbital theory to interpret the properties of conjugated systems, and the methods on which they are based. It does not altogether ignore the important role of computing numbers which can be correlated with experimental observations, but, by and large, it subscribes to Hamming's view that "the purpose of computing is insight, not numbers".[1] The applications are, therefore, somewhat unorthodox compared with computations which seek solely to establish agreement between theoretical predictions, expressed numerically, and experiment. They should appeal to those readers who enjoy finding out the facts for themselves.

The first part of the book, which deals with the Hückel method, contains material which could be adapted for inclusion in undergraduate courses. It is evident, from the few π-electrons problems that can be solved easily (e.g. butadiene, benzene, and possibly naphthalene, where simplifications due to symmetry may be used) and, which therefore qualify for detailed treatment in standard texts on molecular orbital theory, that computational difficulties prevent students from acquiring the kind of practical experience that ensures a good understanding and appreciation of the subject. These difficulties are largely resolved by the use of modern computers, where solutions for molecules containing ten to twenty conjugated atoms are obtained in a few seconds. Thus, although some undergraduates courses are designed primarily to present a statement of the method and its practical use in describing molecular properties, there is a strong case for recommending that, when a computer is available, serious consideration should be given to the experimental approach adopted in this book. This would encourage the student to discover for himself, by carrying out sets of calculations which lie within the region of physical validity, and beyond, if necessary, properties which characterize molecular orbital descriptions, and, in some cases, quantum mechanical concepts on which they are based. Computational methods of this kind can be used to illustrate and explain theoretical principles, to identify questions of interpretation, and generally to enhance the teaching of quantum chemistry.

The later sections of the book, which deal with self-consistent field and configuration interaction calculations, are more relevant to the graduate about to begin research in quantum chemistry, although they could also be considered as an appendix to an undergraduate course. Applications

---

[1] R. W. Hamming, Numerical Methods for Scientists and Engineers, McGraw Hill, New York, 1962.

to π-electron problems offer the simplest, and in some ways the most illuminating practical examples of these methods, which merit careful study since a good understanding of the solutions provides a valuable background for more advanced work. SCF–CI calculations are also important for experimental chemists seeking to use theoretically reliable methods for interpreting results, especially where these involve excited states.

Computer programs written in the FORTRAN II programming language, for solving the secular equations of the Hückel method, the nonlinear equations of the self-consistent field method for π-electron ground states, and configuration interaction equations for excited states, are presented with typical input data and computed results in each case. It has been a major objective to ensure that these programs are easy to use in practice, and special purpose subroutines have been incorporated to accept input data in the simplest possible forms. Thus, an incidence matrix of 0's and 1's which identify non-adjacent and adjacent atoms respectively, defines a framework of conjugated carbon atoms; any conjugated molecule is then represented by simple modifications of elements of the incidence matrix of an appropriate parent carbon framework. Elsewhere, a hexagonal grid is specified, which generates automatically within a subroutine, prescribed atom co-ordinates and molecular integrals required in various parts of the programs. Similarly, a single data record specifies a set of π-electron configurations to be taken into account automatically in calculating excited states. The same simple forms of input data apply to both the Hückel and SCF–CI programs, and it is therefore, easy to turn to the more sophisticated methods, having first gained some experience in using the programs of the Hückel method.

Both sets of programs are designed to operate in core, and to cater for up to 30 conjugated atoms. Under these conditions, the Hückel programs require roughly 12,000 words, and the SCF–CI around 19,000 words of 24 bit core store, with floating point numbers stored in two words. Rather less than one third of the total storage accommodates the object code in each case and the programs can be reduced in size simply by diminishing the dimensions of the arrays. For example, it would be possible to obtain Hückel solutions for molecules comparable in size to naphthalene on a small departmental machine with, say, 5,000 words of core store. Alternatively, the programs are easily segmented. Solutions of the Hückel equations for molecules containing 10 conjugated atoms are obtained in around 1–3 seconds on machines with core speeds of

2–6 microseconds, and corresponding SCF–CI solutions take roughly 10–20 times as long.

In preparing this book I have gained immeasurable help from the texts referenced at the end of Chapter 1, especially Parr's review, with its original papers, and Streitwieser's monograph. Most of the programs presented in the book were originally written at a time when programming implied machine code, and the high speed store consisted of a dozen or so nickel delay lines, and I recall with pleasure the enthusiasm of those days, and the help I received in particular from Dr. Colin Reeves and Mr. Tom Hayward. I owe a similar debt to Professor Roy McWeeny and his research group, and to Mrs. Jean Bishop and Miss Carolyn Healings for patiently preparing the manuscript.

H. H. GREENWOOD

Contents

ix

Contents   **xi**

1

Introduction

The MO method of quantum chemistry has provided, since its introduction around the early 1930's, a versatile framework for the interpretation of the physical and chemical properties of molecules. The scope of the method can be measured by the variety of applications that can be described by the theory at different levels of sophistication, which range from purely qualitative interpretations of molecular properties to large-scale calculations on small molecules, which take into account all terms of the hamiltonian operator, and obtain solutions of increasing accuracy by variational techniques.

This book is concerned exclusively with MO methods devised for the treatment of π-electron systems of conjugated molecules, which are chosen, in the first place, largely for their mathematical simplicity. Thus, a prime advantage of π-electron methods lies in the comparative simplicity of the sets of equations describing the models, and of the methods of solution. In the Hückel scheme, which is the simplest MO model, the equations can, for example, be written down virtually by inspection of the molecular framework, provided appropriate atom and bond parameters are known. Yet the solutions obtained from these unsophisticated versions provide quantitatively consistent accounts of many aspects of aromatic chemistry. Apart from the technical simplicity of formulation and solution, π-electron methods also provide a simple, yet adequate framework for illustrating most of the principles, practice, and descriptive qualities of MO theory, and are well endowed with properties that offer scope for analysis by both theoretical and computational techniques.

However, in spite of the simplicity of formulation, the computational effort involved in solving the equations can become substantial for all but the smallest conjugated molecules. As a result, qualitative methods of interpretation based upon either the classical electronic theories of Ingold and Robinson, or upon the concept of resonance structures proposed by Pauling and Wheland, which provide attractive accounts of molecular properties, are still more widely cultivated and favoured. The situation has, however, changed significantly in recent years through the increasing availability of electronic digital computers, and there are prospects that the quality of the MO description may, as a result, become more widely appreciated. However, the use of computers calls for new and different methods of approach even when solving familiar problems, because of the sheer versatility of machine computations. Numerical calculations which would, otherwise, be directed towards the computation of quantities for comparison with experiment, take on a new meaning, since it becomes possible, on a computer, to experiment with the models themselves. The computer becomes, in effect, like a laboratory tool, by which controlled experiments may be imposed upon theoretical models to study their properties and relevance to the physical systems described.

It should be emphasized at the outset, that in studying computational aspects of π-electron theory, we shall not be concerned with explanations of the quantum mechanical foundations of MO theory, nor with the derivation of equations defining π-electron methods. It will be assumed that the reader is familiar with these methods, or has access to texts which provide the appropriate definitions.[1-5] Computational methods are, after all, mainly concerned with the solution of the equations, with the interpretation of observed molecular properties in terms of the solutions obtained, and with the analysis of theoretical models themselves.

The computer programs provided in later chapters obtain solutions of the sets of equations defining the models, and calculate energy levels and orbitals, charge densities, bond orders, free valences, polarizability coefficients and other quantities that have traditionally been determined in π-electron calculations. However, in choosing material for discussion in later chapters, we recognize that the importance of these calculations for the interpretation of the physical and chemical properties of conjugated molecules has already been widely described elsewhere. Admirable summaries have been given in various books and articles, particularly in Streitwieser's comprehensive review,[1] and it would be pointless to repeat these results here. In consequence, computations of quantities which are intended for direct comparisons with experimental results are, throughout this book, considered quite briefly, though they represent an important area of application, in which the computer is used primarily as a high-speed calculator. Attention is focussed, more often, on the design of experiments which illustrate, within the relatively simple framework of

π-electron theory, the scope of numerical methods in analysing the properties and theoretical implications of proposed models.

The book falls broadly into two parts, the first part being devoted to π-electron theory in the Hückel MO approximation, and the second to SCF–CI methods. The material itself can be identified with certain major landmarks in the development of π-electron theory, namely (a) the introduction of the Hückel method,[6] (b) the perturbation method of Coulson and Longuet-Higgins,[7] (c) Roothaan's SCF method[8] in the approximation proposed by Pariser and Parr,[9] and by Pople,[10] for the treatment of π-electron systems, and (d) CI methods.[11,12] No attempt is made to discuss the various extensions and modifications of these 'standard' methods, which may be significant in particular areas of application, since the conceptual factors associated with each of the main stages of development are more fundamental and relevant to π-electron theory itself, and to the computational techniques employed.

The equations of the Hückel MO method for π-electron systems are stated in Chapter 2. A matrix diagonalization method of solution is explained in some detail, since it represents the core of the computer program for solving the Hückel equations, and is also embodied in programs described later which solve both the SCF and CI problems. Complete listings of the FORTRAN programs are given at the end of this chapter, with input-data specifications, and a printed output of Hückel solutions for a set of chosen molecules.

Chapter 3 deals with certain aspects of the perturbation method, as applied in Hückel theory, which has hitherto proved invaluable in obtaining approximate charge densities, bond orders and similar quantities when parameters are changed in value from those describing a parent hydrocarbon. The discussion does not aim, however, to present a case for using perturbation methods. On the contrary, these methods are largely made obsolete by the use of high-speed computers, since solutions of the Hückel equations for each value of a set of modified parameters, which scan a range covering the values applicable in perturbation theory, and beyond, would consume merely seconds of computer time on a modern machine. The main objective of this chapter is to show that the many analytical properties of the perturbation method applied in Hückel theory are associated with, and ultimately dependent upon, stronger relationships which apply to solutions of the secular equations themselves. These are the relationships normally found in systematically designed computer calculations. Terminology proves somewhat problematic since related solutions of the secular equations are called 'conjugate', a term which appears to be

4   **Computing Methods in Quantum Organic Chemistry**

technically acceptable, but is less felicitous in the context of computations
for conjugated molecules. The terminology is retained, however, since the
meaning is usually clear from the context. The analytical properties are
traced, in outline only, from those that apply to the first and second
coefficients of perturbation formulae, which are used in practical applications, through a generalization to all terms of an expansion formula,
and finally to the finite changes, obtained without truncating the formulae.
Examples illustrating the various relationships are presented in the form
of problems at the end of the chapter.

Various applications of the Hückel method are then considered in
Chapter 4. Traditional calculations of charge densities, bond orders and
similar quantities are, as indicated earlier, discussed quite briefly, and
mainly in the context of mesomeric substitution, where trivial errors can
be made in specifying program-input data. The rest of the chapter is held
together, not by the physics and chemistry of the π-electron systems studied,
but by the contribution each section makes in demonstrating various ways
of applying the programs to obtain solutions which illustrate properties of
the models, and theoretical implications of interpretations which use them.
Thus, the second section considers general properties characterizing energy
level and orbital changes as parameters of the Hückel equations vary
systematically, and shows how certain constraints operate to bring about
a degree of orbital 'localization'. The next two sections serve primarily to
introduce the formulae used in calculating dipole moments and transition
moments for π-electron 'excitations' in the computer programs described
and listed at the end of the chapter. Although the calculation of excitation
energies by the Hückel method is theoretically unreliable, it is possible
to identify, with the aid of computed transition moments, the deficiencies
of the model, and the form of solutions obtained when CI methods are
used. Thus, the Hückel calculations serve to prepare the way, within a
simple theoretical framework, for an understanding of the nature of the
CI problem which is discussed in the final chapter. The final section in this
chapter investigates d_π-p_π π-electron systems, and provides interesting
information relating to the nature of conjugation, which has some bearing
upon the concept of aromaticity as applied to these systems. The discussion
is made possible because the formulation of the problem is, in terms of
input data to the computer program, essentially similar to conventional
p_π-p_π problems.

Chapter 5 discusses MO theories of the chemical reactions of conjugated
molecules. These theories are generally based upon definitions of reactivity
indices associated with models of different stages of the reaction path.

Indices associated with early stages of reaction mechanisms are usually defined in terms of perturbation coefficients describing modified ground-state configurations. The secular equations for modified systems can, however, be solved directly by the computer programs, and changes in energy levels and orbitals, which are not described by the perturbation method, can be recognized. It turns out that anomalies in predictions of active positions, by reactivity indices defined by perturbation methods in certain conjugated molecules, are removed when the complete calculations are made. However, the same computer calculations provide evidence which suggests that certain reactivity indices that correlate numerically, cannot, on conceptual grounds, be associated with the same kind of reaction mechanism. In fact, solutions of the secular equations for modified ground-state configurations, which are easily calculated by computer methods for various ranges of parameter changes, provide evidence of conceptual incompatibilities amongst indices that lie outside the scope of the perturbation methods by which they are defined. Since there has been, hitherto, considerable disagreement concerning the validity of reactivity indices in describing the chemical reactions of conjugated molecules, it is all the more important that changes in energy levels and orbitals associated with perturbation formulae which contain the definitions of reactivity indices should be explored in some detail by computer methods.

The final two chapters of the book are concerned with SCF and CI methods, respectively. Here, as elsewhere, the relevant equations are presented without derivation, though individual terms of the non-linear SCF equations are interpreted in physical terms which, to some extent, compensates for the absence of a derivation, and enables the method to be used effectively within a computational context. Furthermore, the iterative method of solution of the SCF equations, described in Chapter 6, is based upon a matrix-diagonalization process analogous to that used in solving the Hückel equations, and the computer programs can be regarded as extensions of those used earlier. At the same time, input data for both SCF and Hückel programs are virtually identical, and the solutions take the same form in calculating energy levels and orbitals, charge densities, bond orders, free valences and dipole moments. The SCF programs are, therefore, just as easy to use in practice as the Hückel programs. The applications of the SCF programs which are discussed within the text are largely concerned with investigating properties of the non-linear equations. The most important of these concerns the existence of 'conjugate' solutions, with properties virtually identical to those found for Hückel theory. These relationships show that predictions of ground-state properties for heteromolecules derived from parent alternant hydrocarbons (AH) are largely unaffected by the formal neglect, in Hückel theory, of electron-repulsion terms. For example, comparable ground-state charge distributions are obtained for these π-electron systems from both methods of approximation. A further section considers the definition of electronegativity within the SCF description, and associates the concept not only with properties of the atom, but also, significantly, with its site in the molecule.

The final chapter on the CI method leaves more unsaid than elsewhere about the nature of the equations used. Nevertheless, it describes, in detail, numerical forms of the equations, in making comparisons between applications to naphthalene and quinolene, and shows how the differences can be directly related to the solutions obtained. These solutions are described in terms of energies and wavefunctions for excited singlet and triplet states, and computed transition moments. The computer program itself is linked to the SCF programs to form a SCF–CI 'package' which provides a systematic description of π-electron ground and excited states; and, although the calculations are intrinsically much more complicated than those of the Hückel method, the input data for complete sets of SCF–CI solutions are virtually identical with those already used in applying the Hückel programs. No new techniques are, therefore, required to apply the SCF–CI programs, and the experience gained in using the Hückel programs should, therefore, be adequate preparation for adopting the more advanced methods.

Obviously, in a book on computational methods of analysis, the relevance of topics discussed can best be recognized by practical use of the programs. Although theoretical explanations are available, as indicated in several instances, the computational approach provides a simple, direct, and efficient method of gaining an understanding of the properties of models, and of the methods on which they are based. This, in essence, is the main theme of the book, and will be lost to the reader unless the appropriate practical work is pursued.

1.1 REFERENCES

1. A. Streitwieser. *Molecular Orbital Theory for Organic Chemists*, Wiley, New York, 1961.
2. K. Higasi, H. Baba, and A. Rembaum. *Quantum Organic Chemistry*, Wiley, New York, 1965.
3. R. G. Parr. *Quantum Theory of Molecular Electronic Structure*, Benjamin, New York, 1964

4. L. Salem. *The Molecular Orbital Theory of Conjugated Systems*, Benjamin, New York, 1966
5. M. J. S. Dewar. *The Molecular Orbital Theory of Organic Chemistry*, McGraw–Hill, New York, 1969.
6. E. Hückel, *Z. Physik*. **70**, 204, (1931).
7. C. A. Coulson and H. C. Longuet Higgins, *Proc. Roy. Soc.* **A191**, 39 (1947). *Ibid*, **A192**, 16 (1947).
8. C. C. J. Roothaan, *Rev. Mod. Phys.*, **23**, 69 (1951).
9. R. Pariser and R. G. Parr, *J. Chem. Phys.*, **21**, 466 (1953).
10. J. A. Pople, *Trans. Faraday Soc.*, **49**, 1375 (1953).
11. J. A. Pople, *Proc. Phys. Soc.* (*London*), **A68**, 81 (1955).
12. R. Pariser, *J. Chem. Phys.*, **24**, 250 (1956).

2

Hückel Theory

The approximations introduced in formulating the Hückel method for conjugated molecules fall into two categories; those that are concerned with the definition of a π-electron hamiltonian operator, and those that produce simplifications in the secular equations.

The basic approximation of the first kind is that of π–σ separation, which assumes that the π electrons can be treated independently of the σ-bonded framework, except in so far as the framework itself creates an 'effective' field in which the π electrons move. A Schrodinger equation for a system of n π electrons can then be written in the form

$$h_\pi \psi = \epsilon \psi \qquad (2\text{-}1)$$

where the π-electron hamiltonian h_π is given by

$$h_\pi(1, 2, \ldots n) = \sum_{i=1}^{n} (-\tfrac{1}{2}\nabla^2(i) + V_\pi(i)) + \tfrac{1}{2} \sum_{i,j=1}^{n} \frac{1}{r_{ij}} \qquad (2\text{-}2)$$

in which V_π represents the potential energy of a π electron in the field of the σ-bonded framework or 'core'. This can be visualized as a resultant field obtained when all the π electrons are removed to infinity. Next, the independent particle approximation is introduced, in which a π electron is assumed to move in the field of the σ-bonded framework and an averaged field of the remaining π electrons. The operator h_π then simplifies to the form

$$h_\pi(1, 2, 3, \ldots n) = \sum_{i=1}^{n} h_\pi^{\text{eff}}(i) \qquad (2\text{-}3)$$

where

$$h_\pi^{\text{eff}}(i) = -\tfrac{1}{2}\nabla^2(i) + V(i) \qquad (2\text{-}4)$$

is the effective hamiltonian operator for a single π electron, and where $V(i)$ represents the potential energy that incorporates, in some average way, the π-electron-repulsion terms referring to the remaining π electrons, which are included explicitly in equation (2-2). With this approximation the Schrodinger equation (2-1) becomes separable into n-equivalent equations

$$h\psi = \epsilon\psi \tag{2-5}$$

in which h is the operator given in (2-4), and ψ a MO.

Approximate solutions of (2-5) are sought in the form of MOs ψ chosen as linear combinations

$$\psi = \sum_{r=1}^{N} c_r \phi_r \tag{2-6}$$

of the $2p_z$ atomic orbitals ϕ_r associated with each of the N-conjugated atoms. Equation (2-5) can be 'solved' in the sense that a 'best' value for ϵ can be obtained within the framework of the given basic set of orbitals ϕ_r $(r = 1, 2, \ldots N)$ by minimizing the expression for ϵ equivalent to (2-5), namely

$$\epsilon = \frac{\int \psi^* h\psi \, d\tau}{\int \psi^* \psi \, d\tau} \tag{2-7}$$

with respect to variation of the parameters c_r appearing in the expansion (2-6) for the MO ψ. This procedure produces the secular equations, at which stage approximations of the second kind are introduced.

2.1 THE SECULAR EQUATIONS

Minimization of ϵ in (2-7) with respect to variation of the coefficients c_r in (2-6) leads to the set of equations

$$\frac{\partial \epsilon}{\partial c_r} = 0 \qquad (r = 1, 2, 3, \ldots N) \tag{2-8}$$

that are linear in the coefficients c_r, and have the algebraic form

$$\sum_{s=1}^{N} (h_{rs} - \epsilon S_{rs})c_s = 0 \qquad (r = 1, 2, 3, \ldots N) \tag{2-9}$$

where

$$h_{rs} = \int \phi_r^* h\phi_s \, d\tau \tag{2-10}$$

and

$$S_{rs} = \int \phi_r^* \phi_s \, d\tau \tag{2-11}$$

The equations have a non-trivial solution provided the determinant is zero

i.e. $$|h_{rs} - \epsilon S_{rs}| = 0 \qquad (r, s = 1, 2, \ldots N) \tag{2-12}$$

This determinant can be expanded in the form of a polynomial equation

$$\Delta(\epsilon) = 0$$

of degree N in ϵ, and the N roots ϵ_j ($j = 1, 2, \ldots N$) are the energy levels, or eigenvalues of equation (2-5). Substituting a value ϵ_j back into equations (2-9) and solving, gives a set of coefficients c_{rj} ($r = 1, 2, \ldots N$) defining the MO ψ_j or eigenvector in equation (2-5) corresponding to the particular eigenvalue ϵ_j where

$$\psi_j = \sum_{r=1}^{N} c_{rj}\phi_r \qquad (2\text{-}13)$$

Hückel theory now introduces approximations that simplify the equations (2-9) by ignoring all overlap integrals between normalized atomic orbitals, so that

$$S_{rs} = 0 \qquad (r \neq s)$$

$$= 1 \qquad (r = s) \qquad (2\text{-}14)$$

and all terms h_{rs} that do not refer to neighbouring atoms r and s.

i.e. $\qquad\qquad\qquad h_{rs} = 0 \qquad (r, s \text{ non-neighbours}) \qquad (2\text{-}15)$

The remaining h_{rs} terms then fall into two groups,

$$\alpha_r \equiv h_{rr} = \int \phi_r^* h \phi_r \, d\tau \qquad (2\text{-}16)$$

called the 'coulomb' integrals α_r that refer to single atomic orbitals ϕ_r and

$$\beta_{rs} \equiv h_{rs} = \int \phi_r^* h \phi_s \, d\tau \qquad (r, s \text{ neighbours}) \qquad (2\text{-}17)$$

the 'resonance' integrals involving atomic orbitals ϕ_r and ϕ_s on adjacent conjugated atoms. With these approximations, the secular equations reduce to the simpler form

$$(\alpha_r - \epsilon)c_r + \sum_{s=1}^{N} \beta_{rs}c_s = 0 \qquad (s \neq r = 1, 2, \ldots N) \qquad (2\text{-}18)$$

and the secular determinant becomes

$$\Delta(\epsilon) = \begin{vmatrix} (\alpha_1 - \epsilon) & \beta_{12} & \beta_{13} & \cdots\cdots & \beta_{1N} \\ \beta_{21} & (\alpha_2 - \epsilon) & \beta_{23} & \cdots\cdots & \beta_{2N} \\ \beta_{31} & \beta_{32} & (\alpha_3 - \epsilon) & \cdots\cdots & \beta_{3N} \\ \vdots & \vdots & & & \vdots \\ \beta_{N1} & \beta_{N2} & \beta_{N3} & \cdots\cdots & (\alpha_N - \epsilon) \end{vmatrix} = 0 \qquad (2\text{-}19)$$

in which most of the β_{rs} ($= \beta_{sr}$, assuming real orbitals ϕ) are zero, non-zero terms arising only when r and s are neighbouring atoms. Equations (2-18) and (2-19) define the Hückel method in its simplest form; other forms relax the overlap (2-14) and nearest neighbour (2-15) approximations, but these modifications will not be considered.

The MOs ψ_j of equation (2-6) corresponding to the energy levels ϵ_j, obtained as solutions of (2-5), are orthogonal, so that

$$\int \psi_i^* \psi_j \, d\tau = 0$$

This equation expresses an analytical independence of the MOs, and implies that a property, such as the charge distribution, of electrons associated with ψ_i can be computed from ψ_i only. In the case of non-orthogonal orbitals, part of the given property would be computed from ψ_j; hence the value of orthogonal MOs. It is customary to normalize each ψ_i to unity, so that

$$\int \psi_i^* \psi_i \, d\tau = 1$$

which, by substitution from (2-13), assuming real orbitals ϕ_r gives

$$\sum_{r=1}^{N} c_{ri}^2 \int \phi_r^2 \, d\tau + 2 \sum_{r \neq s=1}^{N} c_{ri} c_{si} \int \phi_r \phi_s \, d\tau = 1$$

In Hückel theory, this expression reduces to the form

$$\sum_{r=1}^{N} c_{ri}^2 = 1$$

when the orthonormality condition (2-14) for atomic orbitals is introduced.

The $n \, \pi$ electrons may now be assigned to the MOs ψ_i with spin functions α and β attached. Assume that we are considering a closed-shell ground state in which the lowest M energy levels are doubly occupied, so that $n = 2M$; then a total wavefunction Ψ_0 for the ground state may then be written as the product

$$\Psi_0 = \psi_1(1)\bar{\psi}_1(2)\psi_2(3)\bar{\psi}_2(4) \ldots \psi_i(\mu)\bar{\psi}_i(\mu + 1) \ldots \psi_M(n - 1)\bar{\psi}_M(n) \quad (2\text{-}20)$$

where $\psi \equiv \psi\alpha$, $\bar{\psi} \equiv \psi\beta$ denote spin assignments. Wavefunctions for configurations in which the $n \, \pi$ electrons are assigned in some other way to the available set of MOs ψ_j can be written in a similar form. Consider the 'excited' configuration obtained by transferring an electron from an occupied orbital ψ_i of the ground state to an unoccupied orbital $\psi_{k'}$. Then

$$\Psi_{i \to k'} = \psi_1(1)\bar{\psi}_1(2) \ldots \psi_{k'}(\mu)\bar{\psi}_i(\mu + 1) \ldots \psi_M(n - 1)\psi_M(n)$$

describes the product wavefunction when the spin of the excited electron is unchanged. The energy of a configuration of n π electrons is obtained as the sum over the corresponding occupied energy levels

$$\mathscr{E} = \sum_{j=1}^{N} \nu_j \epsilon_j \qquad (2\text{-}21)$$

where $\nu_j = 0$, 1 or 2 denotes the occupancy of the jth level. This form implies, amongst other matters, that the excited triplet configuration obtained by excitation from ψ_i to $\bar{\psi}_{k'}$ with change of spin, is equal in energy to the corresponding singlet configuration, in the Hückel approximation.

In fact, the product form (2-20) is not an acceptable representation for electron configurations, since it fails to express correctly the indistinguishability of electrons and the Pauli exclusion principle. The most convenient way of fulfilling the principle is to use determinantal wavefunctions as proposed by Slater.[1] However, these forms become significant only when electron repulsion terms are accounted for explicitly (e.g. equation 2-2) in the hamiltonian operator for π electrons, as applied in the SCF (Chapter 6) and CI methods (Chapter 7). In such cases, configuration energies are no longer simple sums (2-21) taken over occupied energy levels.

2.2 MATRIX FORMATION

The secular equations (2-9) or (2-18) when written in matrix form describe a conventional matrix eigenvalue problem. For example, equations (2-18) can be written as

$$\sum_{s=1}^{N} h_{rs} c_s = \epsilon c_r \qquad (r = 1, 2, \ldots N)$$

or

$$(\mathbf{h} - \epsilon\mathbf{I})c = 0 \qquad (2\text{-}22)$$

where $\mathbf{h}$ is a $N \times N$ matrix, with elements defined (equations 2-16 and 17) in the basis of atomic orbitals ϕ_r; c is a $N \times 1$ column vector, with elements that are coefficients in (2-6) of the orbital basis vectors ϕ_r; and ϵ is a scalar. $\mathbf{I}$ is the unit matrix of order N. Suppose, in the first instance, that the solution is known in terms of the eigenvalues ϵ_j ($j = 1, 2, \ldots N$) and the corresponding vectors ψ_j ($j = 1, 2, \ldots N$) of equation (2-13). Then the rules of matrix multiplication show that (2-22) may be generalized to the form

$$\mathbf{hC} = \mathbf{CE} \qquad (2\text{-}23)$$

in which $\mathbf{h}$, $\mathbf{C}$ and $\mathbf{E}$ are all $N \times N$ matrices. $\mathbf{E}$ is a diagonal matrix whose non-zero elements are the eigenvalues ϵ_j, and $\mathbf{C}$ the matrix constructed from the column vectors c_j ($\equiv c_{1j}, c_{2j}, c_{3j} \ldots c_{Nj}$).

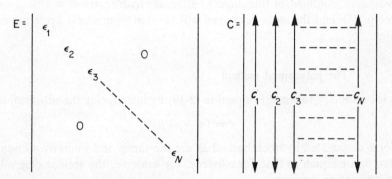

The ϵ_j and c_j ($\equiv \psi_j$) are ordered in the same way in $\mathbf{E}$ and $\mathbf{C}$, though not necessarily in the particular order given above.

Since the eigenvectors c_j are orthonormal

$$\mathbf{C}^\dagger\mathbf{C} = \mathbf{I} \tag{2-24}$$

where $\mathbf{C}^\dagger$ is the transpose of $\mathbf{C}$, and $\mathbf{I}$ is the unit matrix. Therefore, from (2-23)

$$\mathbf{C}^\dagger\mathbf{h}\mathbf{C} = \mathbf{C}^\dagger\mathbf{C}\mathbf{E} = \mathbf{E} \tag{2-25}$$

It follows that solution of the eigenvalue problem consists in finding an orthogonal matrix $\mathbf{C}$ which transforms the given symmetric matrix $\mathbf{h}$ to diagonal form. The diagonal elements will then be the eigenvalues or energy levels, and the columns of $\mathbf{C}$ the corresponding eigenvectors or MOs.

2.3 SOLUTION OF THE SECULAR EQUATIONS

Matrix methods for solving the secular equations of Hückel theory are now used as exclusively on computers as polynomial methods were previously used for hand-machine calculations. The polynomial method can be programmed, but precautions must be made to handle satisfactorily particular computational problems that can arise in practice; degeneracies can, for example, produce such problems, and it may be necessary in certain situations to employ double length working. By comparison, matrix methods are straightforward. However, the polynomial method can be a valuable source of intuitive ideas about the structure of energy-level diagrams, and, to this extent, can provide useful preliminary notions about the nature and origin of certain physical phenomena. A short discussion, confined, at this stage, to alternant hydrocarbons will be given (see p. 44), and the ideas developed will be used in practical applications later.

A. The polynomial method

It is customary to simplify equation (2-19) by introducing the substitution

$$x = (\alpha - \epsilon)/\beta$$

For a conjugated hydrocarbon all αs are the same, and similarly all non-zero βs are equal, so that, for ethylene, for example, the secular determinant equation becomes

$$\Delta(x) = \begin{vmatrix} x & 1 \\ 1 & x \end{vmatrix} = 0$$

$$\text{or} \quad (x^2 - 1) = 0$$

which has the roots $x_1 = -1$ and $x_2 = +1$, or $\epsilon_1 = \alpha + \beta$ and $\epsilon_2 = \alpha - \beta$, with ϵ_1 the lower, since β is negative. Similarly, the equation for butadiene takes the form

$$\Delta(x) = \begin{vmatrix} x & 1 & 0 & 0 \\ 1 & x & 1 & 0 \\ 0 & 1 & x & 1 \\ 0 & 0 & 1 & x \end{vmatrix} = 0$$

or $$x^4 - 3x^2 + 1 = 0$$

with roots

$$x_1 = -(\sqrt{5} + 1)/2 \quad \text{or} \quad \epsilon_1 = \alpha + 1{\cdot}618\beta$$
$$x_2 = -(\sqrt{5} - 1)/2 \quad\quad\quad \epsilon_2 = \alpha + 0{\cdot}618\beta$$
$$x_3 = +(\sqrt{5} - 1)/2 \quad\quad\quad \epsilon_3 = \alpha - 0{\cdot}618\beta$$
$$x_4 = +(\sqrt{5} + 1)/2 \quad\quad\quad \epsilon_4 = \alpha - 1{\cdot}618\beta$$

The corresponding MOs ψ_j ($j = 1, 2, 3, 4$) are obtained by substituting each ϵ_j back, in turn, into the secular equations (2-18) and normalizing It will be noted that the Hückel method does not make a distinction.

between *cis*- and *trans*-butadiene since the non-zero off-diagonal elements arise only between adjacent atom pairs. This simple structure of the secular determinant provides, for chains of conjugated carbon atoms, a recurrence relationship of the form

$$P_N = xP_{N-1} - P_{N-2}$$

where $P_N \equiv P_N(x)$ is the characteristic polynomial for a chain of N atoms. It is interesting to note how the series begins by constructing P_2 for ethylene from P_1 and P_0. Clearly P_1 relates to an isolated carbon atom with determinant

$$P_1(x) \equiv \Delta(x) = |x|$$

and P_0 to the 'bond', or off-diagonal term 1. Then

$$P_2 = xP_1 - P_0 = x^2 - 1$$

For the allyl π-electron system

$$P_3 = xP_2 - P_1 = x^3 - 2x$$

and, for butadiene,

$$P_4 = xP_3 - P_2 = x^4 - 2x^2 - (x^2 - 1)$$
$$= x^4 - 3x^2 + 1$$

Similar recurrence relationships can be obtained for conjugated systems other than chains, though particular problems occur when the construction involves the closure of rings. It is a fairly straightforward matter to build up a 'library' of polynomials for structural residues that will simplify the determination of the characteristic equation for a new conjugated system. The technique has been described in detail by Heilbronner,[2] Streitwieser,[3] and others.

For present purposes, the polynomial formulation is of interest, not primarily as a computational method, but for the insight it provides on the distribution of energy levels in conjugated systems. Consider, for example, the determinant for two conjugated atoms of the same kind

$$\begin{vmatrix} x & \delta \\ \delta & x \end{vmatrix} \equiv (x^2 - \delta^2) = 0$$

$$\text{or} \quad x = \pm\delta$$

with an interaction or 'bonding' term δ. For simplicity, assume the atoms are carbon so that $\delta = 1$ would represent the ethylene system. When δ is zero, the two roots coincide in the zero of energy, and a non-interacting system is represented. As δ is increased from zero the degeneracy is removed, the two levels separate symmetrically and increasingly until the 'ethylene' solution is reached, and beyond, if $\delta > 1$.

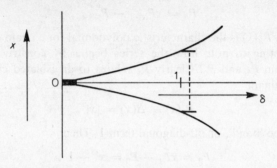

The diagram simply expresses a well-known theoretical property of 'repulsion' between energy levels of interacting systems; the interaction is represented by non-zero off-diagonal terms in the secular equations, and the magnitude of 'repulsion' increases with increase in the interaction term. The levels of the allyl system can, therefore, be visualized as being derived by 'fusion' of ethylene and a single carbon atom, and butadiene as 'fusion' of a carbon atom and allyl.

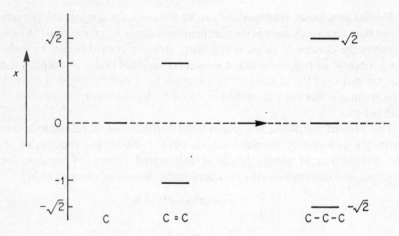

The extension of these constructions to large molecules is usually straightforward, since energy-level diagrams for many molecules are now available in the literature. The polynomial formulation will not, however,

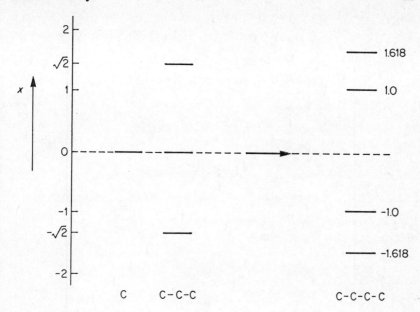

be pursued further for the time being, though it will be invoked subsequently at appropriate places throughout the text. It may, however, be remarked that it is sometimes more valuable to obtain a qualitative understanding of a physical process from comparative studies of energy-level diagrams, than from perturbation formulae, where details of the physical interpretation are seldom directly accessible.

B. Matrix diagonalization

The procedure recommended for solution of the secular equations by computer methods is that of matrix diagonalization as represented in equation (2-25). Of the various techniques that are available, the Jacobi method for real symmetric matrices will be described in outline, and a computer program (JACOBI SCOFI 1 routine) provided.

The method is based upon an iterative technique[4] for annihilating, in turn, off-diagonal elements of the given matrix A by two-dimensional rotations. Assume that the element A_{pq} is to be annihilated, and consider the transformation

$$T'AT = A^* \tag{2-26}$$

where

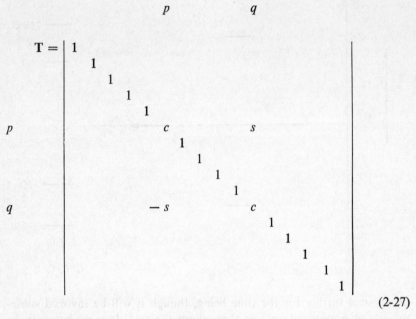

$$(2\text{-}27)$$

in which $c \equiv \cos\theta$, $s \equiv \sin\theta$ all other elements being 1 along the diagonal and 0 otherwise. Then the orthogonal transformation (2-26) represents a rotation in the (p, q) two-dimensional plane through angle θ. The elements of $\mathbf{A}^*$ are the same as those of $\mathbf{A}$ except in the pth and qth rows and columns, where the following values are obtained

$$A_{pp}^* = c^2 A_{pp} + s^2 A_{qq} + 2sc A_{pq}$$
$$A_{qq}^* = s^2 A_{pp} + c^2 A_{qq} - 2sc A_{pq}$$
$$\left.\begin{array}{l} A_{pk}^* = c A_{pk} + s A_{qk} \\ A_{qk}^* = -s A_{pk} + c A_{qk} \end{array}\right] \quad k \neq p, q$$
$$A_{pq}^* = (c^2 - s^2) A_{pq} - sc(A_{pp} - A_{qq}) \tag{2-28}$$

Annihilation of the (pq)th element means that $A_{pq}^* = 0$, which define the required angle of rotation θ by the equation

$$\tan 2\theta = \frac{2A_{pq}}{A_{pp} - A_{qq}} \tag{2-29}$$

In principle rotations should be carried out for every non-zero pivot A_{pq} but since the transformations affect elements other than the pivotal pair, subsequent transformations will undo the annihilation of previous pivots. Clearly, diagonalization will not result from a single pass through the matrix **A** and the method is, therefore, essentially iterative and not finite.

It is easy to show from equations (2-28) that

$$A_{pk}^{*2} + A_{qk}^{*2} = A_{pk}^{2} + A_{qk}^{2}$$

and

$$A_{kp}^{*2} + A_{kq}^{*2} = A_{kp}^{2} + A_{kq}^{2}$$

for all $k \neq p, q$, and, therefore, the sum of squares of off-diagonal elements excluding the (pq)th is invariant under the transformation. This means that the total sum of squares of off-diagonal elements has been reduced by $2A_{pq}^{2}$ and it can be shown, again by application of equations (2-28) that this quantity has been absorbed in the sum of squares of diagonal terms. Successive pivoting therefore transfers the sum of squares of off-diagonal elements on to the diagonal, and the matrix becomes progressively diagonalized.

In practice the matrix is scanned for selected pivots A_{pq}, and rotations are made on each occasion until the matrix converges to diagonal form within a prescribed terminating limit, which might, for example, refer to the magnitude of the largest off-diagonal element. The complete process corresponds to the sequence of operations

$$\ldots T_r' \ldots T_3' T_2' T_1' A T_1 T_2 T_3 \ldots T_r \ldots$$

which converges to the form

$$U'AU = D \tag{2-30}$$

where **D** is diagonal to the prescribed limit, and

$$U = T_1 T_2 T_3 \ldots T_r \ldots T_n \tag{2-31}$$

where the sequence $T_1 T_2 \ldots T_r \ldots T_n$ represents the successive two-dimensional rotations, T_n being the final rotation of the process.

Procedures for selecting pivots and other practical details of the computing technique are discussed in a later section describing the computer programs.

2.4 CHARGE DENSITIES, BOND ORDERS AND FREE VALENCES

Many physical and chemical properties of conjugated molecules are described, not by the MOs themselves, but in terms of certain quantities such as the charge density, bond order and free valence which are derived from the MOs as indicated below.

The probability distribution for a π electron occupying the MO ψ_j is (assuming, for simplicity, real orbitals ϕ_r)

$$\psi_j^2 = [c_{1j}^2\phi_1^2 + c_{2j}^2\phi_2^2 + \cdots + c_{Nj}^2\phi_N^2$$
$$+ 2c_{1j}c_{2j}\phi_1\phi_2 + 2c_{1j}c_{3j}\phi_1\phi_3 + \cdots$$
$$\cdots + 2c_{rj}c_{sj}\phi_r\phi_s + \cdots] \qquad (2\text{-}32)$$

and, since ψ_j is normalized

$$\int \psi_j^2 \, d\tau = \left[\sum_{r=1}^{N} c_{rj}^2 \int \phi_r^2 \, d\tau + 2 \sum_{r<s<1}^{N} c_{rj}c_{sj} \int \phi_r\phi_s \, d\tau \right] = 1 \qquad (2\text{-}33)$$

By the overlap approximation (2-14) the second sum in (2-33) is zero, and the integrated probability distribution for the jth orbital therefore gives

$$\sum_{r=1}^{N} c_{rj}^2 = 1 \qquad (2\text{-}34)$$

Individual terms $\mathrm{e}c_{rj}^2$ of this sum multiplied by the electronic charge, therefore represent (negative) charges associated with individual conjugated atoms r contributed by a π electron occupying the MO ψ_j. It follows that the sum

$$q_r' = \mathrm{e} \sum_{j=1}^{N} \nu_j c_{rj}^2 \qquad (2\text{-}35)$$

taken over the complete set of MOs, where ν_j ($=0$, 1 or 2) is the occupation number, represents the total π-electron charge on atom r. The dimensionless quantity

$$q_r = \sum_{j=1}^{N} \nu_j c_{rj}^2 \qquad (2\text{-}36)$$

is called the π-electron *charge density** at atom r. It is obvious, from the definition, that

$$\sum_{r=1}^{N} q_r = n \qquad (2\text{-}37)$$

where n is the total number of π electrons.

The *bond order* associated with the bond joining atoms r and s is similarly defined as

$$p_{rs} = \sum_{j=1}^{N} \nu_j c_{rj}c_{sj} \qquad (2\text{-}38)$$

---

* 'Charge population' would be a better term for this quantity, but since the description 'charge density' is used almost universally it will be accepted in this text also.

Bond-order terms make no contribution to the charge distribution in the Hückel approximation since the corresponding terms in the probability distribution (2-32) integrate (or sum) to zero when the overlap approximation is applied.

The *free valence* F_r at atom r is defined in terms of the bond orders

$$F_r = N_{\max} - N_r \qquad (2\text{-}39)$$

where N_r is the sum of π bond orders between atom r and its neighbours, and $N_{\max} = \sqrt{3}$ is a theoretical maximum value of N_r derived from trimethylenemethane (see Streitwieser).[3] The equations (2-36, 38 and 39) represent generalized definitions, but for the usual closed-shell ground-state configuration $\nu_j = 2$ for occupied MOs and 0 otherwise.

Now the energy $\mathscr{E}$ of a closed-shell ground state

$$\mathscr{E} = 2\sum_j \epsilon_j \qquad (j = 1, 2, \ldots M) \qquad (2\text{-}40)$$

can be expressed in terms of the charge densities q_r and bond orders p_{rs} by substituting each occupied orbital ψ_j in turn in equation (2-7) to find the corresponding ϵ_j. Thus

$$\epsilon_j = \sum_{r=1}^{N} c_{rj}^2 \alpha_r + 2\sum_{r<s=1}^{N} c_{rj}c_{sj}\beta_{rs} \qquad (2\text{-}41)$$

and, therefore,

$$\mathscr{E} = 2\sum_{j=1}^{M} \epsilon_j = \sum_{r=1}^{N} q_r \alpha_r + 2\sum_{r<s=1}^{N} p_{rs}\beta_{rs} \qquad (2\text{-}42)$$

For conjugated hydrocarbons all coulomb integrals α_r are assumed to be equal ($\alpha_r = \alpha$), and in the simplest approximation β_{rs} is put equal to a uniform value β throughout, and the energy expression (2-42) then simplifies to (using 2-37)

$$\mathscr{E} = n\alpha + 2\beta(\textstyle\sum p_{rs}) \qquad (2\text{-}43)$$

Assume for the moment that all resonance integrals β are formally put equal to zero, so that $\mathscr{E} = n\alpha$ represents the energy of the n π electrons associated with the same set of N carbon atomic $\phi_r(2p_z)$ orbitals in the absence of conjugation. It follows that in the energy expression (2-43) the term $2\beta(\sum p_{rs})$ represents the energy of conjugation, and it is therefore convenient to fix a zero of energy equal to the coulomb integral α for a carbon atom so that $\mathscr{E}$ measures, directly, the conjugation energy.

2

The coulomb integrals for conjugated atoms X other than carbon can now be expressed in the form

$$\alpha_X = \alpha + \delta\alpha_X \qquad (2\text{-}44)$$

and, since π-electron energy is measured in units of β the energy difference $\delta\alpha_X$ can itself be represented in terms of β

$$\delta\alpha_X = h_X\beta \qquad (2\text{-}45)$$

Similarly, resonance integrals other than those between adjacent carbon atoms can also be represented as multiples of β

$$\beta_{XY} = k_{XY}\beta \qquad (2\text{-}46)$$

Both h_X and k_{XY} are dimensionless parameters.

2.5 COMPUTER PROGRAMS WITH LISTINGS

A FORTRAN program for solving Hückel π-electron problems is presented in the form of a set of subprograms, each of which performs a distinct part of the complete calculation; a MAIN program controls the sequence of operations. The complete program is designed to provide processing for a series of molecules, each referred to a parent hydrocarbon which is defined on input by an 'incidence' matrix with elements 1 and 0 specifying respectively neighbouring and non-neighbouring atoms. The number of different molecules 'derived' from the same hydrocarbon parent is specified by the identifier NDER and the total number of parents by NMOLS. A molecule differs from the corresponding parent in one or more matrix elements which represent modifications of coulomb integrals (2-45) and/or resonance integrals (2-46). The number of modifications that brings the form for a parent hydrocarbon to that of a required molecule is specified by the identifier NMOD which appears in subroutine MODH. The individual subprograms can be identified as follows:

MAIN   — the main, control program.
INPT   — reads Hückel semimatrix for a parent hydrocarbon to store.
PAHY   — initiates parent hydrocarbon matrix.
MODH   — modifies selected elements for prescribed molecule.
JACOBI — diagonalization routine.
ORDR   — orders eigenvalues and vectors.
PMAT   — constructs P, the bond order matrix.
FVAL   — constructs F, the vector of free valences.
OTPT   — outputs results.

AIN

&FORTRAN;   TIME = 0000 02

&LIST;   TIME = 0000 02
```
    1*  C      HÜCKEL CALCULATIONS
    2*         DIMENSION A(30,30),ADIAG(30),U(30,30),PRS(30,30),FV(30)
    3*         DIMENSION Z(30)
    4*         READ(7,99)NMOLS
    5*     99  FORMAT(I4)
    6*         DO 10 KMOLS=1,NMOLS
    7*         WRITE(2,100)KMOLS
    8*    100  FORMAT(1H1,13H MOLECULE NO.,I4)
    9*         CALL INPT(N,M,A)
   10*         READ(7,99)NDER
   11*         DO 10 KDER=1,NDER
   12*         CALL PAHY(N,A,ADIAG,Z)
   13*         CALL MODH(N,A,ADIAG,Z)
   14*         EPS=1E-16
   15*         NIT=0
   16*         CALL SCOFI1(N,A,ADIAG,U,NIT,EPS)
   17*         CALL ORDR(N,A,ADIAG,U)
   18*         CALL PMAT(N,M,U,PRS)
   19*         CALL FVAL(N,PRS,FV,A)
   20*         CALL OTPT(N,M,A,ADIAG,U,PRS,FV)
   21*     10  CONTINUE
   22*         STOP
   23*         END
```

INPT

The routine begins by reading N, the number of conjugated atoms, and
M, the number of doubly occupied orbitals, so that 2M is the total number
of π electrons.

Hückel matrices for parent hydrocarbons are read in the form of
'incidence' matrices with off-diagonal elements 1 or 0 indicating neighbouring and non-neighbouring atoms respectively. Diagonal elements of
this matrix are also zero.

The elements are read by columns (a) and stored in an array **NUCK** in
upper semimatrix form

(a) (b)

so that the input semimatrix for benzene is that given in (b). The routine
changes the sign of the 1 off-diagonal elements, following input, to ensure
that bonding terms in a Hückel matrix **HUCK** are negative; the original

incidence matrix is printed for the record, in an equivalent lower semi-matrix form. (See the following subsection on data specification.)

```
 1*          SUBROUTINE INPT(N,M,HUCK)
 2*          DIMENSION HUCK(30,30),NUCK(30,30)
 3*          READ(7,99)N,M
 4*          READ(7,100)((NUCK(I,J),I=1,J),J=1,N)
 5*          DO 17 J=1,N
 6*          DO 17 I=1,J
 7*       17 HUCK(I,J)=-NUCK(I,J)
 8*          DO 16 J=1,N
 9*       16 WRITE (2,101) (NUCK(I,J),I=1,J)
10*      101 FORMAT (40I2)
11*       99 FORMAT(2I3)
12*      100 FORMAT(80I1)
13*          RETURN
14*          END
```

PAHY

As noted in the description of the JACOBI (SCOFI 1) diagonalization routine, processing takes place in the lower semimatrix, and the upper semimatrix is unchanged. This routine brings down into the lower semi-matrix of A the matrix **HUCK** of the parent hydrocarbon which is pre-served in the upper half of A. The diagonal elements are duplicated, for processing purposes, by storing them in the array **ADIAG**. Elements of an array **Z** are also set equal to unity (see Chapter 4).

```
 1*          SUBROUTINE PAHY(N,A,ADIAG,Z)
 2*          DIMENSION A(30,30),ADIAG(30)
 3*          DIMENSION Z(30)
 4*          DO 18 J=1,N
 5*          DO 17 I=1,J
 6*       17 A(J,I)=A(I,J)
 7*          Z(J)=1.0
 8*       18 ADIAG(J)=A(J,J)
 9*          RETURN
10*          END
```

MODH

Elements of the parent hydrocarbon currently stored in the lower semi-matrix **A** and **ADIAG** may be modified in this program. The number of modifications changing the parent to the prescribed molecule is read into the identifier NMOD and each element to be changed is specified by its row and column subscripts I and J, and the value X to be substituted. The program ensures that only the lower semimatrix of A is modified. If I equals J the corresponding element of **ADIAG** is modified. The modified elements for all calculations are then recorded on the lineprinter.

If the eigensolution of the parent hydrocarbon itself is required, the case must be included in specifying the number of variations NDER in the MAIN program, and the modification number NMOD made zero.

```
 1*        SUBROUTINE MODH(N,A,ADIAG,Z)
 2*        DIMENSION A(30,30),ADIAG(30)
 3*·       DIMENSION Z(30)
 4*        READ(7,99)NMOD
 5*        IF(NMOD)18,18,19
 6*    19  WRITE (2,103)
 7*   103  FORMAT (///14H MODIFICATIONS)
 8*        DO 17 K=1,NMOD
 9*        READ(7,100)I,J,X
10*        WRITE (2,102)I,J,X
11*   102  FORMAT(I3,2X,I3,3X,F7.3)
12*        IF (I-J)15,16,14
13*    16  ADIAG(J)=X
14*        GO TO.17
15*    15  A(J,I)=X
16*        GO TO 17
17*    14  A(I,J)=X
18*    17  CONTINUE
19*    99  FORMAT(I4)
20*   100  FORMAT(I2,I2,F6.3)
21*·   18  RETURN
22*        END
```

JACOBI (SCOFI 1)

The rotations (2-26, 27) operate upon the lower semimatrix of the symmetrical Hückel matrix A (or **HUCK**), and the upper half, including the diagonal, is preserved. The diagonal elements of the lower half, that are changed on rotation, are stored separately in the one-dimensional array **ADIAG**.

Various techniques have been proposed and used in practice for selecting pivots A_{pq} for rotation. One method rotates about those pivots that exceed a prescribed threshold value, which itself is systematically reduced when no remaining off-diagonal element exceeds the current threshold; in another method, pivots are chosen as the largest off-diagonal element in a row (or column). A numerical criterion must be prescribed to specify a limit for convergence, and so terminate the iteration process. This could, for example, be a limiting value for the sum of squares of off-diagonal elements, reduced by the order of the matrix. The given program SCOFI 1 adopts a simple, somewhat naïve criterion that, in effect, covers both cases. The terminating criterion for convergence is that the magnitude $|A_{pq}|_{max}$ of the largest off-diagonal element in the final scan shall be less than $1 \cdot 10^{-8}$; it is, in fact, applied in the form

$$(A_{pq})^2 \leqslant \epsilon$$

where ϵ is set in the MAIN program in the identifier EPS. This same number is used for selecting pivots, so that rotation about the pqth element is skipped only when

$$|A_{pq}| < 1 \cdot 10^{-8}$$

In practice rotation takes place about virtually every off-diagonal element during the first three scans; there is then a reduction in the number of rotations in the fourth scan, and the iteration process converges rapidly to meet the required criterion usually after seven scans. With Hückel matrices which are initially sparse, the sum of squares of off-diagonal elements becomes more or less evenly distributed over all elements throughout the first few scans, and it is not obvious that a more refined selection of pivots would save time. Subsequently, selection may be profitable, but at this stage the iteration process converges rapidly in any case. Similar observations apply to the criterion for terminating the process. The more sophisticated criteria require additional calculations, at least during the final scans, and the time for computation could appreciably exceed that required for further scans which rapidly lower the largest off-diagonal element. These remarks, which are based on observations on Hückel matrices of orders 10 to 20, are intended to suggest that increased sophistication should be incorporated with discrimination, and not automatically, since the simple approach may be just as efficient.

The SCOFI 1 routine, as it stands, is short and suitable for machines with a small fast store, and, being simply related to the analysis, is appropriate to the present context. It has a theoretical advantage in producing, automatically, orthonormal eigenvectors over degenerate subspaces, a situation that must often be resolved independently in certain alternative methods, such as the codiagonal matrix methods of Givens and Householder.[5] These degenerate eigenvectors are not, however, generally obtained in forms reflecting the molecular symmetry, which is often a source of degeneracy.

The matrix to be diagonalized, of order N, is stored in lower semi-matrix form below the diagonal in **A** with diagonal elements in **ADIAG**. Normally NIT is set equal to zero on entry. A unit matrix of order N is then constructed in **U** and transformed by the sequence of rotations of equation (2-31) to give the final **U** whose columns are the eigenvectors of **A**. The elements of **U*** obtained from a current **U** matrix by the rotation that annihilates the pqth element of **A** are the same as those of **U** except in the pth and qth columns where the following relationships hold

$$U_{kp}^* = cU_{kp} + sU_{kq}$$

$$U_{kq}^* = -sU_{kp} + cU_{kq}$$

These equations are complementary to those of equations (2-28) which give the transformed elements of **A**.

Hückel Theory

If NIT is not zero on entry, construction of the unit matrix in **U** is bypassed. This can be useful facility when approximate eigenvectors are known and stored on entry in **U**; the technique proves advantageous in solving iteratively the simplified Hartree-Fock SCF equations for π-electron systems described later.

```
 1*         SUBROUTINE SCOFI1(N,A,ADIAG,U,NIT,EPS).
 2*         DIMENSION A(30,30),U(30,30),ADIAG(30)
 3*         IF(NIT)10,999,10
 4*     999 DO 8 J=1,N
 5*         DO 9 I=1,N
 6*       9 U(I,J)=0.0
 7*       8 U(J,J)=1.0
 8*      10 AMAX=0.0
 9*         DO 11 I=2,N
10*         JUP=I-1
11*         DO 11 J=1,JUP
12*         AII=ADIAG(I)
13*         AJJ=ADIAG(J)
14*         AOD=A(I,J)
15*         ASQ=AOD*AOD
16*      28 IF(ASQ-AMAX)23,23,27
17*      27 AMAX=ASQ
18*      23 IF(ASQ-EPS)11,11,12
19*      12 DIFFR=AII-AJJ
20*         IF(DIFFR)13,15,15
21*      13 SIGN=-2.0
22*         DIFFR=-DIFFR
23*         GOTO 16
24*      15 SIGN=2.0
25*      16 TDEN=DIFFR+SQRT(DIFFR*DIFFR+4.0*ASQ)
26*         TANK=SIGN*AOD/TDEN
27*         C=1.0/(SQRT(1.0+TANK*TANK))
28*         S=C*TANK
29*         DO 24 K=1,N
30*         XJ=C*U(K,J)-S*U(K,I)
31*         U(K,I)=S*U(K,J)+C*U(K,I)
32*         U(K,J)=XJ
33*         IF(K-J)17,24,18
34*      17 XJ=C*A(J,K)-S*A(I,K)
35*         A(I,K)=S*A(J,K)+C*A(I,K)
36*         A(J,K)=XJ
37*         GOTO 24
38*      18 IF(K-I)19,24,21
39*      19 XJ=C*A(K,J)-S*A(I,K)
40*         A(I,K)=S*A(K,J)+C*A(I.K)
41*         A(K,J)=XJ
42*         GOTO 24
43*      21 XJ=C*A(K,J)-S*A(K,I)
44*         A(K,I)=S*A(K,J)+C*A(K,I)
45*         A(K,J)=XJ
46*      24 CONTINUE
47*         ADIAG(I)=C*C*AII+S*S*AJJ+2.0*S*C*AOD
48*         ADIAG(J)=C*C*AJJ+S*S*AII-2.0*S*C*AOD
49*         A(I,J)=0
50*      11 CONTINUE
51*         IF(AMAX-EPS)20,20,10
52*      20 RETURN
53*         END
```

ORDR

The eigenvalues and vectors produced by the JACOBI (SCOFI 1) routine are obtained in the same arbitrary sequence. This routine orders the eigenvalues from the lowest to the highest value and the corresponding vectors accordingly. Negative eigenvalues correspond to negative, and, therefore, bonding energy levels; positive eigenvalues represent anti-bonding levels.

```
 1*          SUBROUTINE ORDR(N,A,ADIAG,U)
 2*          DIMENSION A(30,30),ADIAG(30),U(30,30),UTEST(30)
 3*          DO 40 K=1,N
 4*          ATEST=ADIAG(K)
 5*          JTEST=K
 6*          DO 41 J=K,N
 7*          IF(ADIAG(J)-ATEST)42,41,41
 8*       42 ATEST=ADIAG(J)
 9*          JTEST=J
10*       41 CONTINUE
11*          ADIAG(JTEST)=ADIAG(K)
12*          ADIAG(K)=ATEST
13*          DO 40 I=1,N
14*          UTEST(I)=U(I,JTEST)
15*          U(I,JTEST)=U(I,K)
16*       40 U(I,K)=UTEST(I)
17*          RETURN
18*          END
```

PMAT

The calculation of charge densities and bond orders involves summations over the lowest M-occupied orbitals, appropriately ordered in the previous routine.

The program calculates complete bond-order matrices with elements referring to all atom pairs. Diagonal elements are the charge densities

$$q_r \equiv P(IR, IR)$$

and the conventional bond-order terms p_{rs} are those values $P(IR, IS)$ that refer to adjacent atoms. The remaining matrix elements are formal bond orders between non-neighbours.

```
 1*          SUBROUTINE PMAT(N,M,C,PRS)
 2*          DIMENSION C(30,30),PRS(30,30)
 3*          DO 18 IR=1,N
 4*          DO 18 IS=1,IR
 5*          SUM=0.
 6*          DO 17 J=1,M
 7*       17 SUM=SUM+C(IR,J)*C(IS,J)
 8*       18 PRS(IR,IS)=2.0*SUM
 9*          RETURN
10*          END
```

FVAL

The upper semimatrix of A is scanned to identify neighbours s of a prescribed atom r. The corresponding bond orders p_{rs} are summed, and the free valence of atom r is determined from

$$F_r = \sqrt{3} - \sum_s p_{rs} \qquad (r, s \text{ neighbours})$$

```
 1*          SUBROUTINE FVAL(N,PRS,FV,A)
 2*          DIMENSION PRS(30,30),FV(30),A(30,30)
 3*          DO 18 J=1,N
 4*          XNR=0.
 5*          DO 27 I=1,N
 6*          IF(I-J)16,27,28

 7*       16 IF(A(I,J)+0.1)29,29,27
 8*       29 XNR=XNR+PRS(J,I)
 9*          GO TO 27
10*       28 IF(A(J,I)+0.1)39,39,27
11*       39 XNR=XNR+PRS(I,J)
12*       27 CONTINUE
13*       18 FV(J)=1.732   -XNR
14*          RETURN
15*          END
```

OTPT

This subroutine provides a printed layout for a lineprinter with a width
of at least 120 characters. The energy levels, tabulated MOs, total π-
electron energy, charge densities, and free valences are all printed with
appropriate labels attached. The bond-order matrix is printed without
labelling in the semimatrix sequence

$$P_{11} \quad P_{12} \quad P_{22} \quad P_{13} \quad P_{23} \quad P_{33} \quad P_{14} \quad P_{24} \ldots \text{etc.}$$

```
 1*          SUBROUTINE OTPT(N,M,A,ADIAG,U,PRS,FV)
 2*          DIMENSION A(30,30),ADIAG(30),U(30,30),PRS(30,30),FV(30)
 3*          IND=0
 4*          L=0
 5*       16 L=L+1.
 6*          IF(10*L-N)17,18,18
 7*       18 NUPP=N-10*(L-1)
 8*          IND=1
 9*          GO TO 77
10*       17 NUPP=10
11*       77 WRITE(2,101)
12*      101 FORMAT(/14H ENERGY LEVELS)
13*          KLOW=10*(L-1)+1
14*          KUPP=NUPP+10*(L-1)
15*          WRITE(2,99)(K,K=KLOW,KUPP)
16*       99 FORMAT(3X,3H J=,3X,I3,8(7X,I3))
17*          WRITE (2,100)(ADIAG(K),K=KLOW,KUPP)
18*      100 FORMAT(/4X,8F10.4)
19*          WRITE (2,102)   .
20*      102 FORMAT(/16H HÜCKEL ORBITALS)
21*          WRITE(2,99)(K,K=KLOW,KUPP)
22*          DO 27 I=1,N
23*       27 WRITE(2,103)I,(U(I,K),K=KLOW,KUPP)
24*      103 FORMAT(I4,8F10.6)
25*          IF(IND)16,16,15
26*       15 SUM=0
27*          DO 115 J=1,M
28*      115 SUM=SUM+ADIAG(J)
29*          SUM=2*SUM
30*          WRITE(2,105)SUM
31*      105 FORMAT(/27H TOTAL PI-ELECTRON ENERGY =,F10.4)
32*          WRITE(2,104)
33*      104 FORMAT(/17H CHARGE DENSITIES)
34*          WRITE(2,100)((PRS(I,J),I=J,J),J=1,N)
35*          WRITE(2,106)
36*      106 FORMAT(/14H FREE VALENCES)
37*          WRITE(2,100)(FV(I),I=1,N)
38*          WRITE(2,108)
39*      108 FORMAT(/18H BOND-ORDER MATRIX)
40*          WRITE(2,100)((PRS(I,J),J=1,I),I=1,N)
41*          RETURN
42*          END
```

A. Data specification

The following data list provides an input to the Hückel program for benzene, the three diazines, s-triazine, styrene and the hyperconjugation of toluene, the systems to be treated consecutively. It will be observed that the 'incidence' matrix is presented in lower semimatrix form which is convenient for the printed page; provided the matrix elements are punched by rows the data will be read correctly.

0002	NMOLS = number of parent hydrocarbons. (I4)
006003	N, M — in INPT (2I3)

```
1 | 0
2 | 1 0
3 | 0 1 0
4 | 0 0 1 0
5 | 0 0 0 1 0
6 | 1 0 0 0 1 0
```
NUCK (80I1)

benzene

0006	NDER — six variations (I4)
0000	NMOD — benzene (I4)
0001	NMOD — pyridine (I4)
0101–0·500	I, J, X (I2, I2, F6·3)
0002	NMOD — pyridazine
0101–0·500⎤ 0202–0·500⎦	I, J, X
0002	NMOD — pyrimidine
0101–0·500⎤ 0303–0·500⎦	I, J, X
0002	NMOD — pyrazine
0101–0·500⎤ 0404–0·500⎦	I, J, X
0003	NMOD — s-triazine
0101–0·500⎤ 0303–0·500 0505–0·500⎦	I, J, X
008004	N, M — in INPT

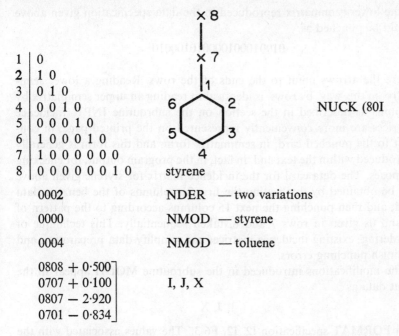

1	0
2	1 0
3	0 1 0
4	0 0 1 0
5	0 0 0 1 0
6	1 0 0 0 1 0
7	1 0 0 0 0 0
8	0 0 0 0 0 0 1 0

NUCK (80I

styrene

0002   NDER   — two variations

0000   NMOD   — styrene

0004   NMOD   — toluene

$$\left.\begin{array}{c} 0808 + 0{\cdot}500 \\ 0707 + 0{\cdot}100 \\ 0807 - 2{\cdot}920 \\ 0701 - 0{\cdot}834 \end{array}\right] \quad \text{I, J, X}$$

The parameters for the toluene calculation were selected from those proposed by I'Haya[6] in discussing hyperconjugation between the methyl group and the ring. The numbering scheme is represented as follows

so that in equations (2-45 and 2-46)

$$h_7 = -0{\cdot}1; \quad h_8 = -0{\cdot}5; \quad k_{78} = 2{\cdot}920; \quad k_{17} = 0{\cdot}834$$

the signs being chosen to match β which is negative, and represented by -1 in the **HUCK** matrices.

Some further comments are needed to clarify certain details of the input-data specification.

Incidence matrices are punched on data cards as strings of 1s and 0s according to the FORMAT specification 80I1, that is, with one character per card column. Thus, the incidence matrix for benzene corresponding

to the lower semimatrix reproduced in the data specification given above
would be punched as

$$010010001000010100010$$
$$\uparrow \;\uparrow \;\uparrow \;\uparrow \quad\; \uparrow \qquad\; \uparrow$$

where the arrows point to the ends of the rows. Reading a lower semimatrix in this way, by rows, is identical to reading an upper semimatrix by
columns as described in the section on the subroutine INPT. Incidence
matrices are more conveniently represented on the printed page, in contrast to the punched card, in semimatrix form, and this form is generally
reproduced within the text and, in fact, in the program output, for checking
purposes. The data card for the incidence matrix for styrene given above
can be obtained by duplicating the first 21 columns of the benzene data
card, and then punching the next 15 columns according to the pattern of
1s and 0s given in rows 7 and 8, taken sequentially. This technique of
'bordering' existing incidence matrices can simplify data preparation and
diminish punching errors.

The modifications introduced in the subroutine MODH appear in the
input data as

$$I, J, X$$

with FORMAT specification I2, I2, F6·3. The values associated with the
subscripts I and J identify the row and column of the element of the
matrix **HUCK** to be modified. If I = J, then X is to be identified with
$\delta\alpha_X$ in equation (2-44) and not with the parameter h_X in (2-45). However,
$\delta\alpha_X$ is to be measured in units of $|\beta|$, so that, in fact, $\delta\alpha_X$ and h_X are equal
in magnitude but opposite in sign. This representation ensures that $\delta\alpha_X = X$
carries the correct sign on an electronegativity basis relative to carbon.
In the data specification for the azines given above, the nitrogen atoms are
correctly represented as negative with respect to carbon, although,
according to equation (2-45) h_X would, correspondingly, be given by $+0·5$.
Similarly, when I $\neq$ J the value of X is to be identified with β_{XY} in equation
(2-46) and not with the parameter k_{XY}. Thus X must always be negative
in this context since it represents a resonance integral, whereas k_{XY} would,
correspondingly, be positive.

The notation used for X avoids a switching of signs, and corresponding
ambiguities which are sometimes associated with the use of the parameters h_X and k_{XY}. A similar feature appears in the form of the program
output where negative eigenvalues imply negative, and, therefore, bonding
energy levels, and positive eigenvalues imply antibonding levels; each
eigenvalue is measured in units of $|\beta|$. Most of the ambiguities which often

arise in practice due to the fact that a chosen unit of energy, β, is negative are, thereby, systematically avoided. The only possible source of uncertainty lies in the natural tendency to interpret off-diagonal 1s of the incidence matrices as representing, more or less directly, the resonance integral β. It is better to interpret the notation literally, so that incidence matrices simply identify neighbouring and non-neighbouring conjugated atoms, and to assume that the programs then construct a representation of the Hückel equations from this data. The fact that this construction simply involves changing the signs of all 1s of the stored incidence matrix **NUCK** in producing a Hückel matrix does not imply a genuine source of ambiguity.

B. Computed results

The following tabulated results were obtained from the data specification given above.

```
   MOLECULE NO.   1
0
1 0
0 1 0
0 0 1 0
0 0 0 1 0
1 0 0 0 1 0

ENERGY LEVELS
   J=    1          2          3          4          5          6

      -2.0000    -1.0000    -1.0000     1.0000     1.0000     2.0000

HÜCKEL ORBITALS
   J=    1          2          3          4          5          6
   1  0.408248  0.183671 -0.547356 -0.561289 -0.135234 -0.408248
   2  0.408248 -0.382189 -0.432742  0.397761 -0.418473  0.408248
   3  0.408248 -0.565859  0.114614  0.163528  0.553707 -0.408248
   4  0.408248 -0.183671  0.547356 -0.561289 -0.135234  0.408248
   5  0.408248  0.382189  0.432742  0.397761 -0.418473 -0.408248
   6  0.408248  0.565859 -0.114614  0.163528  0.553707  0.408248

TOTAL PI-ELECTRON ENERGY =   -8.0000

CHARGE DENSITIES

      1.0000     1.0000     1.0000     1.0000     1.0000     1.0000

FREE VALENCES

      0.3987     0.3987     0.3987     0.3987     0.3987     0.3987

BOND-ORDER MATRIX

      1.0000     0.6667     1.0000    -0.0000     0.6667     1.0000    -0.3333    -0.0000

      0.6667     1.0000     0.0000    -0.3333    -0.0000     0.6667     1.0000     0.6667

      0.0000    -0.3333    -0.0000     0.6667     1.0000


MODIFICATIONS
  1    1    -0.500
ENERGY LEVELS
   J=    1          2          3          4          5          6

      -2.1074    -1.1672    -1.0000     0.8410     1.0000     1.9337

HÜCKEL ORBITALS
   J=    1          2          3          4          5          6
   1  0.520706 -0.571374  0.000000  0.545913  0.000000 -0.323073
   2  0.418504 -0.190609 -0.500000 -0.366024 -0.500000  0.393128
   3  0.361268  0.348897 -0.500000 -0.238101  0.500000 -0.437110
   4  0.342849  0.597839  0.000000  0.566258 -0.000000  0.452102
   5  0.361268  0.348897  0.500000 -0.238101 -0.500000 -0.437110
   6  0.418504 -0.190609  0.500000 -0.366024  0.500000  0.393128

TOTAL PI-ELECTRON ENERGY =   -8.5493

CHARGE DENSITIES

      1.1952     0.9230     1.0045     0.9499     1.0045     0.9230

FREE VALENCES

      0.4247     0.4090     0.3977     0.4022     0.3977     0.4090

BOND-ORDER MATRIX

      1.1952     0.6537     0.9230    -0.0225     0.6694     1.0045    -0.3261     0.0591

      0.6649     0.9499    -0.0225    -0.3306     0.0045     0.6649     1.0045     0.6537

     -0.0770    -0.3306     0.0591     0.6694     0.9230


MODIFICATIONS
  1    1    -0.500
  2    2    -0.500
```

ENERGY LEVELS
```
  J=      1         2         3         4         5         6

      -2.2168   -1.2129   -1.1007    0.7275    0.9298    1.8733
```
HÜCKEL ORBITALS
```
  J=      1         2         3         4         5         6
  1   0.518353 -0.415224  0.347399  0.534799 -0.242692 -0.305457
  2   0.518353 -0.415224 -0.347399 -0.534799 -0.242692  0.305457
  3   0.371572  0.119199 -0.556094  0.121641  0.589683 -0.419479
  4   0.305360  0.559804 -0.264714  0.446311 -0.305573  0.480347
  5   0.305360  0.559804  0.264714 -0.446311 -0.305573 -0.480347
  6   0.371572  0.119199  0.556094 -0.121641  0.589683  0.419479
```
TOTAL PI-ELECTRON ENERGY =   -9.0610

CHARGE DENSITIES
```
      1.1236    1.1236    0.9230    0.9534    0.9534    0.9230
```
FREE VALENCES
```
      0.4186    0.4186    0.4046    0.4041    0.4041    0.4046
```
BOND-ORDER MATRIX
```
      1.1236    0.6408    1.1236   -0.1002    0.6726    0.9230   -0.3322    0.0356

      0.6548    0.9534    0.0356   -0.3322    0.0660    0.6731    0.9534    0.6726

     -0.1002   -0.3139    0.0660    0.6548    0.9230
```


MODIFICATIONS
```
  1   1    -0.500
  3   3    -0.500
```
ENERGY LEVELS
```
  J=      1         2         3         4         5         6

      -2.1928   -1.2808   -1.0767    0.7808    0.9121    1.8575
```
HÜCKEL ORBITALS
```
  J=      1         2         3         4         5         6
  1   0.468386 -0.557345 -0.263607 -0.435162  0.302073 -0.346233
  2   0.427196 -0.000000 -0.489648  0.000000 -0.662398  0.372796
  3   0.468386  0.557345 -0.263607  0.435162  0.302074 -0.346233
  4   0.365705  0.435162  0.337621 -0.557345  0.235852  0.443448
  5   0.333545  0.000000  0.627130  0.000000 -0.517185 -0.477468
  6   0.365705 -0.435162  0.337621  0.557345  0.235852  0.443448
```
TOTAL PI-ELECTRON ENERGY =   -9.1007

CHARGE DENSITIES
```
      1.1990    0.8445    1.1990    0.8742    1.0091    0.8742
```
FREE VALENCES
```
      0.4240    0.4153    0.4240    0.4149    0.3972    0.4149
```
BOND-ORDER MATRIX
```
      1.1990    0.6583    0.8445   -0.0435    0.6583    1.1990   -0.3205   -0.0182

      0.6497    0.8742   -0.0182   -0.3292   -0.0182    0.6674    1.0091    0.6497

     -0.0182   -0.3205    0.1167    0.6674    0.8742
```


MODIFICATIONS
```
  1   1    -0.500
  4   4    -0.500
```
ENERGY LEVELS
```
  J=      1         2         3         4         5         6

      -2.1861   -1.3508   -1.0000    0.6861    1.0000    1.8508
```

HÜCKEL ORBITALS
```
 J=      1          2          3          4          5          6
 1   0.454401  -0.605913  -0.000000  -0.541774  -0.000000  -0.364513
 2   0.383092  -0.257750  -0.500000   0.321310  -0.500000   0.428445
 3   0.383092   0.257750  -0.500000   0.321310   0.500000  -0.428445
 4   0.454401   0.605913  -0.000000  -0.541774  -0.000000   0.364513
 5   0.383092   0.257750   0.500000   0.321310  -0.500000  -0.428445
 6   0.383092  -0.257750   0.500000   0.321310   0.500000   0.428445
```

TOTAL PI-ELECTRON ENERGY =   -9.0738

CHARGE DENSITIES

```
      1.1472    0.9264    0.9264    1.1472    0.9264    0.9264
```

FREE VALENCES

```
      0.4110    0.4108    0.4108    0.4110    0.4108    0.4108
```

BOND-ORDER MATRIX

```
   1.1472    0.6605    0.9264    0.0358    0.6606    0.9264   -0.3213    0.0358

   0.6605    1.1472    0.0358   -0.3394   -0.0736    0.6605    0.9264    0.6605

  -0.0736   -0.3394    0.0358    0.6606    0.9264
```


MODIFICATIONS
```
 1    1    -0.500
 3    3    -0.500
 5    5    -0.500
```

ENERGY LEVELS
```
  J=      1          2          3          4          5          6

     -2.2656   -1.2808   -1.2808    0.7808    0.7808    1.7656
```

HÜCKEL ORBITALS
```
 J=      1          2          3          4          5          6
 1   0.432827   0.005262  -0.643546   0.467923  -0.183130  -0.382092
 2   0.382092  -0.433093  -0.254791  -0.502777  -0.401738   0.432827
 3   0.432827  -0.559958   0.317215  -0.075366   0.496798  -0.382092
 4   0.382092  -0.004109   0.502465   0.599304  -0.234549   0.432827
 5   0.432827   0.554696   0.326330  -0.392557  -0.313668  -0.382092
 6   0.382092   0.437202  -0.247674  -0.096527   0.636287   0.432827
```

TOTAL PI-ELECTRON ENERGY =   -9.6542

CHARGE DENSITIES

```
      1.2030    0.7970    1.2030    0.7970    1.2030    0.7970
```

FREE VALENCES

```
      0.4237    0.4237    0.4237    0.4237    0.4237    0.4237
```

BOND-ORDER MATRIX

```
   1.2030    0.6541    0.7970   -0.0395    0.6541    1.2030   -0.3160    0.0395

   0.6541    0.7970   -0.0395   -0.3160   -0.0395    0.6541    1.2030    0.6541

   0.0395   -0.3160    0.0395    0.6541    0.7970
```

MOLECULE NO.   2
```
0
1 0
0 1 0
0 0 1 0
0 0 0 1 0
1 0 0 0 1 0
1 0 0 0 0 0
0 0 0 0 0 1 0
```

ENERGY LEVELS
```
  J=      1          2          3          4          5          6          7          8

     -2.1358   -1.4142   -1.0000   -0.6622    0.6622    1.0000    1.4142    2.1358
```

HÜCKEL ORBITALS

J=	1	2	3	4	5	6	7	8
1	0.513120	-0.353553	-0.000000	-0.334227	0.334227	-0.000000	-0.353553	-0.513120
2	0.394103	0.000000	-0.500000	-0.307706	-0.307706	-0.500000	-0.000000	0.394103
3	0.328596	0.353553	-0.500000	0.130478	-0.130478	0.500000	0.353553	-0.328596
4	0.307706	0.500000	-0.000000	0.394103	0.394103	0.000000	-0.500000	0.307706
5	0.328596	0.353553	0.500000	0.130478	-0.130478	-0.500000	0.353553	-0.328596
6	0.394103	0.000000	0.500000	-0.307706	-0.307706	0.500000	-0.000000	0.394103
7	0.307706	-0.500000	-0.000000	0.394103	0.394103	0.000000	0.500000	0.307706
8	0.144072	-0.353553	0.000000	0.595183	-0.595183	-0.000000	-0.353553	-0.144072

TOTAL PI-ELECTRON ENERGY = -10.4243

CHARGE DENSITIES

1:0000	1.0000	1.0000	1.0000	1.0000	1.0000	1.0000	1.0000

FREE VALENCES

0:1058	0.4432	0.3947	0.4148	0.3947	0.4432	0.4148	0.8207

BOND-ORDER MATRIX

1:0000	0.6101	1.0000	0.0000	0.6787	1.0000	-0.3012	-0.0000
0:6586	1.0000	0.0000	-0.3213	-0.0000	0.6586	1.0000	0.6101
0:0000	-0.3213	-0.0000	0.6787	1.0000	0.4059	0.0000	-0.0485
0:0000	-0.0485	0.0000	1.0000	0.0000	-0.2527	-0.0000	0.2042
-0:0000	-0.2527	0.9113	1.0000				

MODIFICATIONS

8	8	0.500
7	7	0.100
8	7	-2.920
7	1	-0.834

ENERGY LEVELS

J=	1	2	3	4	5	6	7	8
	-2:8208	-1.9363	-1.0000	-0.9524	0.9861	1.0000	1.9728	3.3506

HÜCKEL ORBITALS

J=	1	2	3	4	5	6	7	8
1	0.296875	0.313683	-0.000000	-0.557649	-0.567200	0.000000	0.369068	-0.211112
2	0.126475	0.378234	-0.500000	-0.305788	0.291506	-0.500000	-0.396821	0.070667
3	0.059889	0.418686	-0.500000	0.266432	0.279742	0.500000	0.413781	-0.025663
4	0.042462	0.432463	0.000000	0.559524	-0.567364	-0.000000	-0.419486	0.015319
5	0.059889	0.418686	0.500000	0.266432	0.279742	-0.500000	0.413781	-0.025663
6	0.126475	0.378234	0.500000	-0.305788	0.291506	0.500000	-0.396821	0.070667
7	0.700824	-0.178762	0.000000	0.096521	-0.028405	-0.000000	0.078592	0.678668
8	0.616232	-0.214254	0.000000	0.194059	0.170621	-0.000000	-0.155819	-0.695200

TOTAL PI-ELECTRON ENERGY = -13.4189

CHARGE DENSITIES

0:9950	1.0051	0.9997	1.0038	0.9997	1.0051	1.0649	0.9266

FREE VALENCES

0:2288	0.4096	0.3977	0.4013	0.3977	0.4096	0.5579	0.7542

BOND-ORDER MATRIX

0:9950	0.6534	1.0051	0.0011	0.6689	0.9997	-0.3275	-0.0043
0:6654	1.0038	0.0011	-0.3311	-0.0003	0.6654	0.9997	0.6534
0:0051	-0.3311	-0.0043	0.6689	1.0051	0.1963	-0.0170	-0.0143
0:0129	-0.0143	-0.0170	1.0649	.0.0150	-0.1249	-0.0022	0.0842
-0:0022	-0.1249	0.9778	0.9266				

END

&END;
TIME = 0001 34
A

2.6 PROBLEMS

1. Construct a NUCK matrix for naphthalene

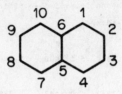

by bordering that given in Section 2.5.A for benzene.

Apply in turn the following modifications

   (i) 0001   NMOD
      0101 − 0·500   I, J, X
  (ii) 0001   NMOD
      0202 − 0·500   I, J, X

to compute solutions for quinolene and isoquinolene. Note the corresponding changes in charge densities, bond orders, free valences and energy levels.

2. Border the NUCK matrix for naphthalene to give that for anthracene and apply similar modifications at symmetrically distinct atoms.

3. Repeat the same procedure in constructing data for phenanthrene and its various N derivatives.

4. Compute Hückel solutions for linear polyenes C_nH_{n+2}, taking $n = 2, 4, 6, 8, 10$; observe bond order, free valence and energy level changes in the series.

5. Compute Hückel solutions for the corresponding cyclic polyenes.

6. Construct a NUCK matrix for azulene

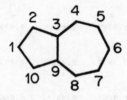

and introduce modifications at various peripheral atoms to represent N substitution. Note that the charge densities in the parent hydrocarbon are not unity.

7. Construct a NUCK matrix for the benzyl framework.

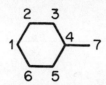

Note that the following two specifications for N, M

  (i) 007004   N, M   (8π electrons)

 (ii) 007003   N, M   (6π electrons)

compute π-electron configurations for (i) the anion and (ii) the cation.
How can the configuration for the radical be determined?

2.7 REFERENCES

1. J. C. Slater, *Phys. Rev.*, **34**, 1293 (1929).
2. E. Heilbronner, *Helv. Chim. Acta.*, **36**, 170 (1953).
3. A. Streitwieser, *Molecular Orbital Theory for Organic Chemists*, Wiley, New York, 1961.
4. J. Greenstadt, *Mathematical Methods for Digital Computers* (Ed. A. Ralston and H. Wilf), Wiley, New York, 1960, Vol. 1, Chap. 7.
5. J. Ortega, *Mathematical Methods for Digital Computers*, Wiley, New York, 1966, Vol. 2, Chap. 4.
6. Y. I'Haya, *Bull. Chem. Soc. Japan*, **28**, 369 (1955).

3

Parametric Properties of Hückel Equations

The methods discussed in the previous chapter determine solutions of the secular equations for any conjugated molecule described by an appropriate set of α, β values. Hückel theory has, however, progressed beyond this stage, notably by the use of perturbation methods, which provide practical advantages in obtaining approximate solutions, and in comparative studies of physical and chemical properties of conjugated molecules.

A typical application is the derivation of approximate solutions for conjugated molecules in which the secular equations differ from those of a parent hydrocarbon in particular coulomb or resonance integrals. The formal representation of pyridine, for example, may differ from that of benzene in the coulomb integral representing the nitrogen atom, which, according to (2-44, 45) may be written in the form

$$\alpha_N = \alpha + \delta\alpha_N$$

with

$$\delta\alpha_N = h_N\beta$$

Since β is negative, the dimensionless, adjustable parameter h_N is positive, ensuring that the nitrogen atom is more electronegative than the carbon atom it replaces. A solution for pyridine can be expressed in terms of finite changes ΔA from the known solution of the parent hydrocarbon

$$\Delta A = \left(\frac{\partial A}{\partial \alpha_N}\right) \delta\alpha_N + \tfrac{1}{2}\left(\frac{\partial^2 A}{\partial \alpha_N^2}\right) \delta\alpha_N^2 + \cdots \qquad (3\text{-}1)$$

where A may represent $\mathscr{E}$, the π-electron energy, q_r, a charge density, p_{st}, a bond order, or any other quantity relating to the description of the solution. The leading partial derivatives in the expansion (3-1) can be determined once and for all, from the solution for the parent hydrocarbon, and thereafter, approximate solutions for pyridine can be found from the change δA obtained by truncating the expression (3-1) for the finite change ΔA after the first or second term; clearly any number of such 'solutions' are obtained by substituting directly prescribed values of $\delta \alpha_N$.

The expansion form (3-1) is more generally written in terms of a change $\delta \alpha_u$ ($\equiv \delta \alpha_N$) in coulomb integral at the uth atom position in a conjugated system, and typical examples of truncated formulae are

$$\delta \mathscr{E} = q_u \delta \alpha_u + \tfrac{1}{2} \pi_{u,u} \delta \alpha_u^2 \tag{3-2}$$

$$\delta q_r = \pi_{r,u} \delta \alpha_u \tag{3-3}$$

$$\delta p_{st} = \pi_{st,u} \delta \alpha_u \tag{3-4}$$

where the symbol δA is reserved for approximate values of the finite change ΔA.

The first term in (3-2) is obtained from equation (2-42), where

$$q_u = \left(\frac{\partial \mathscr{E}}{\partial \alpha_u} \right) \tag{3-5}$$

and the terms $\pi_{r,u}$ and $\pi_{st,u}$ that are known as atom–atom and bond–atom polarizabilities respectively, are, specifically, the first-order derivatives

$$\pi_{r,u} = \left(\frac{\partial q_r}{\partial \alpha_u} \right) \tag{3-6}$$

$$\pi_{st,u} = \left(\frac{\partial p_{st}}{\partial \alpha_u} \right) \tag{3-7}$$

which are independent of $\delta \alpha_u$ and can be calculated from a known solution for the parent hydrocarbon.

The corresponding formulae for a change $\delta \beta_{uv}$ in the resonance integral referring to adjacent atoms u and v are

$$p_{uv} = \tfrac{1}{2} \left(\frac{\partial \mathscr{E}}{\partial \beta_{uv}} \right) \tag{3-8}$$

$$\pi_{r,uv} = \left(\frac{\partial q_r}{\partial \beta_{uv}} \right) \tag{3-9}$$

$$\pi_{st,uv} = \left(\frac{\partial p_{st}}{\partial \beta_{uv}} \right) \tag{3-10}$$

where $\pi_{r,uv}$ and $\pi_{st,uv}$ are the atom–bond and bond–bond polarizabilities respectively.[1,2]

This chapter is devoted to the analytical treatment of parameter variations in Hückel equations, initially within the context of perturbation theory. The objective is not to prepare for the computation of perturbation coefficients; on the contrary, the availability of computer programs for calculating 'exact'* solutions of the Hückel equation means that the approximate solutions obtained by perturbation methods for π-electron systems can largely be dispensed with. Instead, the aim is to present, initially through known properties of polarizability coefficients, important analytical relationships that largely condition the nature of solutions obtained by the Hückel method. Clearly, the computer programs can be used without reference to these properties, and numerical results can, correspondingly, be obtained for comparison with experimental observations. However, both Hückel and SCF methods abound with analytical relationships that should not be ignored, even in the context of numerical calculations which can be carried out rapidly and with negligible personal effort, on a computer. At the lowest level, recognition of these properties will assist in the economic planning of computer runs and, thereby, conserve valuable computer time. More important still, they will provide the investigator with an analytical framework through which valid solutions can be chosen judiciously, and possibly with evidence of the extent to which correlations with experimental results may be parameter-independent.

Most of the properties discussed in this chapter can be verified numerically by appropriate applications of the computer programs described previously, as indicated in the problems presented at the end of this chapter. Calculations of this kind provide elementary demonstrations of the value of computer methods for studying analytical relationships, and large sections of later chapters are similarly concerned to illustrate the use of computers for investigating the properties of theoretical models. Such applications are generally more interesting, instructive, and rewarding than those aimed solely at obtaining numerical correlations between theory and experiment.

The results obtained by the different analytical methods described in this chapter are quoted briefly, and generally without proofs; for this

---

* The term 'exact' implies, in this context, a value of ΔA obtained by direct solution of the secular equations; it is to be distinguished from approximate values δA obtained from truncated forms, e.g. (3-2 to 4) of equation (3-1) by perturbation methods. Direct solutions do not give 'exact' values in the full sense of the term, since the accuracy is limited by numerical procedures, by arithmetical rounding, and by the word length of the machine.

reason, it may be worth identifying the successive stages in the development of the subject matter. Section 3-1 states properties of solutions for AHs derived from the 'pairing' theorem. Section 3-2 presents expressions for various polarizability coefficients, which, as indicated above, are first and second-order terms in the expansion formula (3-1) for ΔA. These coefficients can be expressed either in a conventional form of perturbation theory as certain sums taken over orbitals and energy levels of the unperturbed system, or in an integral form introduced by Coulson and Longuet-Higgins. The integral forms are particularly valuable from an analytical point of view, because they may be developed to include every term of the infinite expansion (3-1), from which may be deduced the analytical properties of ΔA. No attempt is made within the text to explain the derivation of high-order terms, and the technique is presented in outline only for the sake of completeness, and to preserve continuity of the argument, since the principles involved in extending the treatment to the infinite expansion are not too difficult to follow. One further reason for including the integral form here is that similar ideas are later used in Chapter 5 in establishing relationships between reactivity indices. However, in both areas of application, the essential features illustrating the analytical properties under consideration can be established numerically, and rapidly by computer calculations, without reference to the integral forms. The reader may, therefore, be inclined to omit the section dealing with the integral formulation at a first reading, and to reconsider these theoretical matters retrospectively in the light of experience gained by examining the analytical properties numerically. An experimental approach of this kind is advocated throughout the text, which frequently seeks to show how computational methods may be used to illustrate analytical properties of models and methods. Section 3.3 finally demonstrates that all the properties may be deduced directly from the secular equations, firstly in terms of the 'pairing' theorem, and secondly in terms of a generalized pairing of conjugate solutions, defined by related variations $\pm\delta\alpha_u$; formal proofs are outlined, since they do not appear to have been given in the literature.

It is intended that this chapter should convey the idea that systematically planned sets of calculations may be used, not merely to illustrate, but sometimes to detect, properties of a theoretical method. The fact that many properties of the secular equations of Hückel theory have already been discovered does not diminish the argument. In the case of the SCF method for conjugated molecules (see Chapter 6) the non-linearity of the equations makes the derivation and detection of analytical properties rather more difficult, and computer calculations may provide the simplest and most direct method for finding them.

3.1 PROPERTIES OF THE SECULAR EQUATIONS

A set of secular equations (2-18) with dimensions N prescribed, can be regarded as defined in a parameter space of α, β values, so that a characteristic polynomial, its roots, eigenvectors and all derived quantities are completely specified by prescribing a particular set of $[\alpha, \beta]$ values defining a unique point in the parameter space. The dimensions of the equations and of the determinant (2-19) are, of course, fixed by the number of atomic orbitals participating in conjugation, that is, by the value of N in the expansions (2-6 or 13) for ψ. It is possible, therefore, to define the kth partial derivative with respect to any α_u or β_{vw} at a point P in the space specified by a prescribed set $[\alpha, \beta]_P$ of parameter values, and to write

$$\Delta A = \left(\frac{\partial A}{\partial x}\right)_P \delta x + \tfrac{1}{2}\left(\frac{\partial^2 A}{\partial x^2}\right)_P \delta x^2 + \cdots + \frac{1}{k!}\left(\frac{\partial_k A}{\partial x^k}\right)_P \delta x^k + \cdots \quad (3\text{-}11)$$

where $x = \alpha_u$ or β_{vw}, and $A = E$, ϵ_j, c_{sj}, q_r, p_{st}, F_r, and so on. The first and second derivatives of charge densities q_r and bond orders p_{st} are the polarizability coefficients defined by Coulson and Longuet-Higgins,[1,2] which are to be interpreted simply as slopes and curvatures contributing to a finite change ΔA.

It turns out that a prescribed set of α, β values in which all coulomb integrals α are equal, and all resonance integrals β are the same, and for which the topological arrangement of conjugated atoms is 'alternant', gives rise to solutions that possess unique analytical properties. An alternant molecule is, by definition, one in which the starring of alternate neighbours cannot give rise to adjacent starred atoms; thus naphthalene is alternant and azulene is non-alternant.

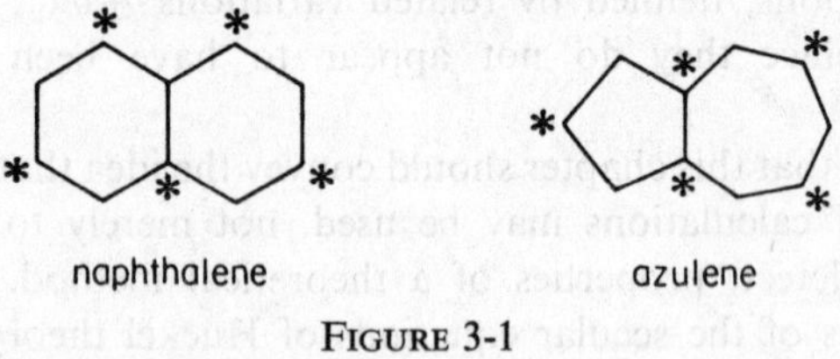

naphthalene   azulene

FIGURE 3-1

A conjugated hydrocarbon that is alternant meets the specified conditions, since, in the simplest approximation, the coulomb integrals for all carbon atoms are the same, and resonance integrals for all carbon–carbon bonds are equal. The unique properties of alternant hydrocarbons (AH) are well known, and can be briefly summarized as follows

(i) Bonding and antibonding energy levels are 'paired', with levels ϵ_j and ϵ_{N-j+1} lying symmetrically above and below the zero α of energy respectively.

(ii) If the orbital ψ_j corresponding to ϵ_j is written in the form

$$\psi_j = \sum_r^* c_{rj}\phi_r + \sum_r c_{rj}\phi_r \qquad (3\text{-}12)$$

where the first sum $\sum^*$ is taken over conjugated atoms that are 'starred', and the second $\sum$ over those that are not starred, then

$$\psi_{j'} = \sum_r^* c_{rj'}\phi_r - \sum_r c_{rj'}\phi_r \qquad (3\text{-}13)$$

is the MO corresponding to $\epsilon_{j'}$ ($j' = N - j + 1$).

(iii) When the number of conjugated atoms is odd, a non-bonding MO (NBMO) of energy $\epsilon_{\mathrm{NBMO}} = \alpha = 0$ is obtained, and

$$\psi_{j=\mathrm{NBMO}} = \sum_r^* b_{rj}\phi_r \qquad (3\text{-}14)$$

since the coefficients of unstarred atoms are zero, provided the starring process is applied so that the number of starred atoms exceeds the unstarred by one.

(iv) When the MOs are filled, in ascending order, by the n π electrons in spin-coupled pairs, the π-electron density q_r at each atom in the neutral molecule is unity.

These are well-known properties, and can readily be confirmed by applying the computer programs described in Chapter 2 to any arbitrarily chosen AH.

3.2 PROPERTIES OF POLARIZABILITY COEFFICIENTS

In considering the analytical properties of polarizability coefficients for AH, the expressions used in practice will be quoted without proof, the reader being referred to the original papers of Coulson and Longuet-Higgins.[1,2] The expressions were originally developed in two forms; firstly by conventional perturbation methods, in terms of the energy levels ϵ_j and MOs ψ_j of the parent hydrocarbon, and, secondly, as contour integrals in which the integrands are functions of the secular determinant and its minors.

In the perturbation method, wavefunctions for the modified π-electron system are expressed in the form of expansions taken over the MOs of the (unperturbed) alternant parent hydrocarbon. As a result, polarizability coefficients are obtained in terms of the coefficients c_{sj} of the atomic orbitals ϕ_s in the MOs ψ_j of the parent AH. Typical first-order coefficients, for example, take the form

$$\pi_{r,u} = 4\sum_{j=1}^{M} \sum_{k=M+1}^{N} \frac{c_{rj}c_{uj}c_{rk}c_{uk}}{\epsilon_j - \epsilon_k} \tag{3-15}$$

$$\pi_{st,u} = 2\sum_{j=1}^{M} \sum_{k=M+1}^{N} \frac{c_{uj}c_{uk}(c_{sj}c_{tk} + c_{tj}c_{sk})}{\epsilon_j - \epsilon_k} \tag{3-16}$$

$$\pi_{st,uv} = 2\sum_{j=1}^{M} \sum_{k=M+1}^{N} \frac{(c_{sj}c_{tk} + c_{tj}c_{sk})(c_{uj}c_{vk} + c_{uk}c_{vj})}{\epsilon_j - \epsilon_k} \tag{3-17}$$

where $M = N/2$ in an AH containing an even number N of conjugated atoms. Using the relationships between paired orbitals ψ_j and $\psi_{j'}$ ($j' = N - j + 1$) in AH, it is not difficult to show that $\pi_{st,u} = 0$.[2] The second-order terms become increasingly complicated, and correct results are obtained only when normalization terms are incorporated.[3] First-order and the more cumbersome second-order polarizability coefficients are readily programmed, but the method appears to hold no prospect either for computing or for studying the properties of higher order terms.

Turning now to the contour-integration method, the following formulae for the charge density q_r and the bond order p_{st} represent typical starting points of the method[1]

$$q_r = 1 - \frac{1}{\pi} \int \left[\frac{\Delta_{r,r}(iy)}{\Delta'(iy)}\right] dy \tag{3-18}$$

$$p_{st} = (-)^{s+t+1} \frac{1}{\pi} \int \left[\frac{\Delta_{s,t}(iy)}{\Delta'(iy)}\right] dy \tag{3-19}$$

Integration is along the y axis in the complex plane, from $-\infty$ to $+\infty$ and (iy) is the argument in all secular determinants and minors. $\Delta_{a,b}$ is the determinant obtained by crossing out the ath row and bth column of Δ in the expression (2-19) and Δ' is the derivative with respect to the argument (iy).

The polarizability coefficients can now be derived[1] by differentiation of the forms (3-18) and (3-19) to give first-order terms

$$\pi_{r,u} = \frac{\partial q_r}{\partial \alpha_u} = \frac{1}{\pi} \int \left[\frac{\Delta_{r,u}}{\Delta} \right]^2 dy \tag{3-20}$$

$$\pi_{st,u} = \frac{\partial p_{st}}{\partial \alpha_u} = \frac{1}{\pi} \int \left[\frac{\Delta_{s,u}\Delta_{t,u}}{\Delta^2} \right] dy \tag{3-21}$$

Similar expressions can be obtained when differentiations are made with respect to β_{vw} but these will not be considered here.

Interesting analytical properties arise when the integrands refer to minors for AH. Consider an even AH, with $N = 2M$ conjugated atoms; $\Delta(\epsilon)$ is then a polynomial of order N, with even powers of ϵ only, in which successive terms alternate in sign, reflecting the symmetric disposition of roots about the origin, $\alpha = 0$, already described. When the appropriate minors have been obtained, it is found that the integrands obtained from the substitution $\epsilon \rightarrow (iy)$ are either real functions of y, or imaginary functions that are odd in (iy) and therefore integrate to zero. The theory can, in fact, be immediately developed further to apply these properties to individual terms in the finite change:

$$\Delta A = \sum_{k=1}^{\infty} \frac{1}{k!} \left(\frac{\partial^k A}{\partial \alpha_u^k} \right) (\delta\alpha_u)^k \tag{3-22}$$

Successive differentiation under the integral sign yields the following formulae[4]

$$\Delta q_r = \sum_{k=1}^{\infty} (-1)^k \frac{1}{\pi} \int \left[\left(\frac{\Delta_{r,u}}{\Delta} \right)^2 \left(\frac{\Delta_{u,u}}{\Delta} \right)^{k-1} dy \right] (\delta\alpha_u)^k \tag{3-23}$$

$$\Delta p_{st} = \sum_{k=1}^{\infty} (-1)^k \frac{1}{\pi} \int \left[\left(\frac{\Delta_{s,u}\Delta_{t,u}}{\Delta} \right) \left(\frac{\Delta_{u,u}}{\Delta} \right)^{k-1} dy \right] (\delta\alpha_u)^k \tag{3-24}$$

in which atom–atom and bond–atom polarizabilities are the leading coefficients corresponding to $k = 1$. The properties of the minors of the secular determinant for AH may now be used to show that alternate integrals in (3-23) and (3-24) are zero,[4] so that the integrands of successive non-zero terms in each expansion differ by the multiplier

$$\left[\frac{\Delta_{u,u}(iy)}{\Delta(iy)} \right]^2 \tag{3-25}$$

The zero terms in (3-23) correspond to even powers, and in (3-24) to odd powers of $\delta\alpha_u$ so that Δq_r is an odd function, and Δp_{st} an even function of $\delta\alpha_u$, which shows why the first and second-order terms of perturbation theory

$$\pi_{st,u} = \left(\frac{\partial p_{st}}{\partial\alpha_u}\right) \quad \text{and} \quad \left(\frac{\partial^2 q_r}{\partial\alpha_u^2}\right)$$

are both zero.

It is interesting to note that, having established these properties, it is then possible to sum the expansions (3-23) and (3-24) in the closed forms given by Fukui *et al.*[5]

$$\Delta q_r = \frac{\delta\alpha_u}{\pi} \int \left(\frac{\Delta_{r,u}}{\Delta}\right)^2 \left[\frac{1}{1 + y^2 G_u^2 \delta\alpha_u^2}\right] dy \tag{3-26}$$

$$\Delta p_{st} = -\frac{\delta\alpha_u}{\pi} \int \left(\frac{\Delta_{s,u}\Delta_{t,u}}{\Delta^2}\right) \left[\frac{y^2 G_u}{1 + y^2 G_u^2 \delta\alpha_u^2}\right] dy \tag{3-27}$$

where

$$G_u = \frac{\Delta_{u,u}(iy)}{(iy)\Delta(iy)} \tag{3-28}$$

is of the form

$$\frac{\text{polynomial in } y^2 \text{ of order } (M-1)}{\text{polynomial in } y^2 \text{ of order } M}$$

G_u is related to the multiplier (3-25) and correspondence with equations (3-23) and (3-24) is easily established by binomial expansion. The analogous formula for the energy is

$$\Delta\mathscr{E} = \delta\alpha_u - \frac{1}{\pi} \int \ln[1 + (\delta\alpha_u)^2 y^2 G_u^2] \, dy \tag{3-29}$$

Coulson and Jacobs[6] have shown how integrals in equations (3-20) and (3-21) may be evaluated numerically using standard quadrature formulae. The polynomial forms $\Delta_{a,b}(\epsilon)$ must, in general, be derived algebraically preceding integration, and the method is, therefore, not ideally suited to automatic computation. Nevertheless, coefficients of any order in (3-23) and (3-24) and finite changes ΔA for prescribed values of $\delta\alpha_u$ in (3-26) and (3-27) can, in principle, be evaluated numerically though it is always much simpler on a computer to calculate ΔA directly from solution of the appropriate secular equations.

3.3 PAIRING PROPERTIES OF CONJUGATE SOLUTIONS

The analytical relationships quoted above—and others—which have been discussed in outline, and without proofs, can in fact be deduced directly from the secular equations; proofs appear to have been given only in terms of corresponding theorems for the π-electron SCF equations,[7] which are more complicated than the Hückel secular equations. The proofs which are outlined below, show that the pairing of levels and orbitals for AH is a particular case of a more general situation, involving a pairing between 'conjugate' solutions, which in the simplest case are obtained from changes $\pm\delta\alpha_u$ in the coulomb integral at atom u. The following relations hold between two 'conjugate' solutions.

(i) Bonding and antibonding levels are 'paired' so that

$$\epsilon_j\left(-\delta\alpha_u\right) = -\epsilon_{N-j+1}\left(+\delta\alpha_u\right) \tag{3-30}$$

(ii) If the orbital ψ_j corresponding to ϵ_j is written in the form

$$\psi_j\left(-\delta\alpha_u\right) = \overset{*}{\sum_r}c_{rj}\phi_r + \sum_r c_{rj}\phi_r \tag{3-31}$$

where $\sum^*$ and $\sum$ denote sums over starred and unstarred atoms respectively, then

$$\psi_{N-j+1}\left(+\delta\alpha_u\right) = \overset{*}{\sum_r}c_{rj}\phi_r - \sum_r c_{rj}\phi_r \tag{3-32}$$

The relations between the levels $\epsilon_j\left(-\delta\alpha_u\right)$ and $\epsilon_{j'}\left(+\delta\alpha_u\right)$ where $j' = N - j + 1$ may be recognized in the energy-level diagram of Figure 3-2, which also includes the levels of the AH (for which $\delta\alpha_u = 0$), that are symmetric about the chosen origin $\alpha = 0$.

The relations (i) and (ii) can be obtained directly from the secular equations

$$-\epsilon c_r + \sum\beta c_s = 0 \quad (r \neq u; r, s \text{ neighbours}; r, s = 1, 2, \ldots N)$$

$$(-\delta\alpha_u - \epsilon)c_u + \sum\beta c_s = 0 \quad (u, s \text{ neighbours}) \tag{3-33}$$

Let ϵ_j be an eigenvalue with $\psi_j \equiv (c_{1j}, c_{2j}, \ldots c_{Nj})$ the corresponding eigenvector. Now change the sign of $\delta\alpha_u$, and change the signs of coefficients in ψ_j of the unstarred set to give a vector $\psi_{j'}$. It is easy to see on inspection that the secular equations are satisfied on substituting $\epsilon_{j'}$ and

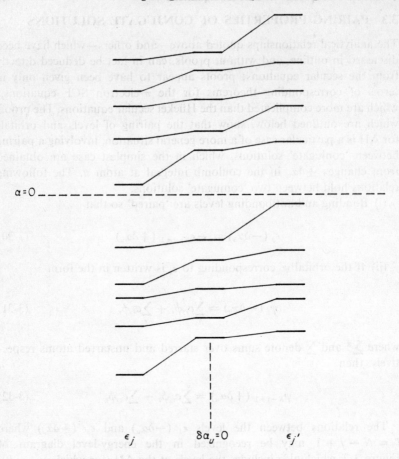

modified energy levels : N=10

FIGURE 3-2

ψ_j, since, in any single equation, the neighbours r and s belong to opposite sets, and the equation changes sign in every term or remains unchanged, so establishing the results (i) and (ii).

The main consequences of the 'pairing' properties are derived most simply from the orthogonality properties of the matrix **C** which collects the N MOs ψ_j ($j = 1, 2, \ldots N$) as column vectors, each element c_{rj} being a coefficient in the atomic orbital basis ϕ_r

$$\psi_j = \sum_r c_{rj}\phi_r \qquad (r = 1, 2, \ldots N)$$

The orthogonality relationships

$$C'C = I \tag{3-34}$$

can be written in terms of orthogonality between columns

$$\sum_{r=1}^{N} c'_{jr} c_{rk} = \delta_{jk} \tag{3-35}$$

and between rows

$$\sum_{j=1}^{N} c_{sj} c'_{jt} = \delta_{st} \tag{3-36}$$

where
$$\delta_{\lambda v} = 1 \quad (\lambda = v)$$
$$\phantom{\delta_{\lambda v}} = 0 \quad (\lambda \neq v)$$

Equations (3-25) represents the orthonormality of MOs, the sum being taken over all atoms r, and (3-26) relates to the bond order matrix where the sum refers to a pair of atoms, s and t, and is taken over all orbitals. The second sum can be separated into two parts

$$\sum_{j=1}^{M} c_{sj} c'_{jt} + \sum_{j=M+1}^{N} c_{sj} c'_{jt} = \delta_{st} \tag{3-37}$$

where the first sum is taken over the $M = N/2$ occupied orbitals, and the second over the unoccupied, or virtual orbitals. Taking the case $s = t$ first, equation (3-37) becomes

$$\tfrac{1}{2} q_s + \tfrac{1}{2} q_s^v = 1 \tag{3-38}$$

where q_s is the charge density defined (2-36) at atom s and q_s^v is a similar sum taken over antibonding, or virtual, orbitals. For AHs, these sums contain squares of coefficients that are equal but may be opposite in sign, and therefore

$$q_s = q_s^v = 1 \tag{3-39}$$

which proves that the charge density is unity for all atoms in an AH. Applying a similar argument to the case $s \neq t$ gives

$$\tfrac{1}{2} p_{st} + \tfrac{1}{2} p_{st}^v = 0 \tag{3-40}$$

where p_{st} is the bond order between any pair of atoms s and t and p_{st}^v is a corresponding sum taken over the virtual orbitals. In AHs the products $c_{sj}c_{jt}'$ for atoms belonging to the starred set carry the same sign in both sums, and equation (3-40) can be satisfied only by the result

$$p_{st} = p_{st}^v = 0 \qquad (s, t \text{ same set}) \qquad (3\text{-}41)$$

However, when s and t belong one to each set, then

$$p_{st} = -p_{st}^v \qquad (s, t \text{ opposite sets}) \qquad (3\text{-}42)$$

satisfies the relationship. The results (3-39, 41, 42) for AHs represent generalizations applicable to the bond-order matrix, of the property that the charge density at all atoms is unity.

The same orthogonality relations (3-35) and (3-36) can be used to establish relations between 'conjugate' solutions for perturbed systems ($\delta\alpha_u \neq 0$). The proofs are straightforward, depending upon the relations between the antibonding, or unoccupied, levels and orbitals of one solution, and the bonding, or occupied, levels and orbitals of its conjugate. The orthogonality relation (3-36) applied to the case $s = t$ gives

$$\tfrac{1}{2}q_s\left(-\delta\alpha_u\right) + \tfrac{1}{2}q_s^v\left(-\delta\alpha_u\right) = 1$$

or

$$\tfrac{1}{2}[1 + \Delta q_s\left(-\delta\alpha_u\right)] + \tfrac{1}{2}[1 + \Delta q_s^v\left(-\delta\alpha_u\right)] = 1$$

and, therefore,

$$\Delta q_s\left(-\delta\alpha_u\right) + \Delta q_s^v\left(-\delta\alpha_u\right) = 0$$

But since squared terms only are involved in defining q_s,

$$q_s^v\left(-\delta\alpha_u\right) = q_s\left(+\delta\alpha_u\right)$$

by equations (3-31) and (3-32), and, therefore,

$$\Delta q_s^v\left(-\delta\alpha_u\right) = \Delta q_s\left(+\delta\alpha_u\right)$$

so that

$$\Delta q_s\left(-\delta\alpha_u\right) = -\Delta q_s\left(+\delta\alpha_u\right) \qquad (3\text{-}43)$$

and, therefore, Δq_s is an odd function of $\delta\alpha_u$.

Applying the orthogonality relation (3-36) to the case $s \neq t$ gives

$$\tfrac{1}{2} p_{st} \left(-\delta\alpha_u\right) + \tfrac{1}{2} p_{st}^v \left(-\delta\alpha_u\right) = 0$$

Now the coefficients of antibonding orbitals $\psi_j \left(-\delta\alpha_u\right)$ differ from the bonding orbitals $\psi_{j'} \left(+\delta\alpha_u\right)$ where $j' = N - j + 1$ in the signs of coefficients of opposite sets, so that

$$p_{st}^v \left(-\delta\alpha_u\right) = +p_{st} \left(+\delta\alpha_u\right) \qquad (s, t \text{ same set})$$

and

$$p_{st}^v \left(-\delta\alpha_u\right) = -p_{st} \left(+\delta\alpha_u\right) \qquad (s, t \text{ opposite set})$$

and, therefore, finally

$$p_{st} \left(-\delta\alpha_u\right) = -p_{st} \left(+\delta\alpha_u\right) \qquad (s, t \text{ same set}) \qquad (3\text{-}44)$$

$$= +p_{st} \left(+\delta\alpha_u\right) \qquad (s, t \text{ opposite sets}) \qquad (3\text{-}45)$$

Since for the parent AH (equation 3-41) p_{st} is zero for s and t in the same set, and the value of p_{st} for s and t in opposite sets is the same for perturbations $\pm\delta\alpha_u$, these last two equations can be written in the alternative form

$$\Delta p_{st} \left(-\delta\alpha_u\right) = -\Delta p_{st} \left(+\delta\alpha_u\right) \qquad (s, t \text{ same set}) \qquad (3\text{-}46)$$

$$= +\Delta p_{st} \left(+\delta\alpha_u\right) \qquad (s, t \text{ opposite sets}) \qquad (3\text{-}47)$$

The last result contains the particular case of the bond order referring to adjacent pairs of conjugated atoms, proving that the change in this bond order is an even function of $\delta\alpha_u$, which in turn contains the first-order result $\pi_{st,u} = 0$. Indeed all the properties of first and second-order theory are included in the theory of finite changes ΔA which, based upon the secular equations themselves, goes much further by presenting simultaneously the properties of all elements of the bond-order matrix, and of the energy levels and orbitals. Moreover these ideas can be extended to include different modifications $\delta\alpha_u$ at different atoms u, and the existence of 'conjugate' solutions in such cases can be established.

Perhaps the simplest way of recognizing these properties in practice is to obtain computer solutions for paired changes $\pm\delta\alpha_u$ applied to atom u of some parent AH and to compare numerical values of the same bond orders and charge densities in both calculations. It should then be possible to recognize immediately equal and opposite changes Δq_s at all atoms s given by equation (3-43) and the bond-order relationships (3-46) and (3-47) for all pairs of atoms s and t and not merely neighbours. Having witnessed these correspondences, details of the essential features of the proofs given in this section should be clarified.

The following data set provides an input to the Hückel program of Chapter 2 that gives two pairs of conjugate solutions obtained with the changes $\delta\alpha_u = \pm 1\cdot0\beta$, $\pm 2\cdot0\beta$ applied at position $u = 1$ in the given naphthalene framework; the solution for the parent hydrocarbon is obtained from the first calculation, NMOD $= 0$.

0001		NMOLS
010005		N, M

```
 1  0
 2  1 0
 3  0 1 0
 4  0 0 1 0
 5  0 0 0 0 0
 6  0 0 0 0 1 0
 7  0 0 0 0 0 1 0
 8  0 0 0 0 0 0 1 0
 9  1 0 0 0 0 0 0 1 0
10  0 0 0 1 1 0 0 0 1 0
```

0005	NDER
0000	NMOD
0001	NMOD
0101 $-$ 1·000	
0001	NMOD
0101 $+$ 1·000	
0001	NMOD
0101 $-$ 2·000	
0001	NMOD
0101 $+$ 2·000	

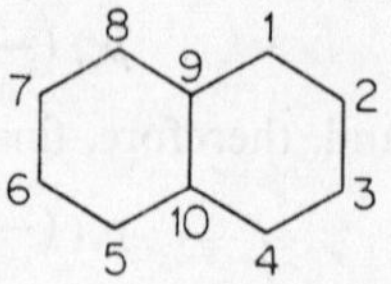

FIGURE 3-3

3.4   COMPUTER PROGRAMS

The subroutine ATAT computes atom–atom polarizabilities from the perturbation formula

$$\pi_{r,u} = 4 \sum_{j=1}^{M} \sum_{k=M+1}^{N} \frac{c_{rj}c_{uj}c_{rk}c_{uk}}{\epsilon_j - \epsilon_k} \tag{3-15}$$

where N is the number of MO levels and orbitals, and M is the number of doubly occupied orbitals of the ground-state configuration. The values are calculated in semimatrix form and stored in full matrix form in the two-dimensional array **PIRU** with subscripts *IR, IU* corresponding to *r* and *u* respectively. The subroutine contains its own output statements, which print the polarizabilities $\pi_{r,u}$ between each atom *u* and all other atoms *r*. Considerable duplication occurs in molecules with symmetry, both in the calculations and in the printed output. In naphthalene, for example, there are only three symmetrically distinct atoms, namely 1, 2 and 9, and the complete set of $\pi_{r,u}$ could be computed by selecting this subset for *u* (or *IU*) and running *r* (or *IR*) through the complete set $r = 1, 2 \ldots 10$. This procedure implies either that parts of ATAT should be rewritten for each new parent hydrocarbon, or that an additional array containing selected values of *u* should be introduced as part of the data, and accessed within the subroutine.

A. Computed results

The polarizabilities $\pi_{r,u}$ for naphthalene[8] as obtained from the subroutine ATAT are presented in Table 3.1. It will be observed that the signs alternate in passing from *u* through successive atoms *r*, and that, for comparable distances of separation, polarizabilities referring to atoms *r* and *u* of opposite sets are appreciably greater than those referring to the same set.

```
 MOLECULE NO.   1
0
1 0
0 1 0
0 0 1 0
0 0 0 0 0
0 0 0 0 1 0
0 0 0 0 0 1 0
0 0 0 0 0 0 1 0
1 0 0 0 0 0 0 1 0
0 0 0 1 1 0 0 0 1 0
```

```
ATOM-ATOM POLARIZABILITIES

ATOM      1
-0.4428   0.2134  -0.0177   0.1394   0.0232  -0.0064   0.0323  -0.0267   0.0889  -0.0036

ATOM      2
 0.2134  -0.4049   0.1096  -0.0177  -0.0064   0.0326  -0.0001   0.0323  -0.0073   0.0486

ATOM      3
-0.0177   0.1096  -0.4049   0.2134   0.0323  -0.0001   0.0326  -0.0064   0.0486  -0.0073

ATOM      4
 0.1394  -0.0177   0.2134  -0.4428  -0.0267   0.0323  -0.0064   0.0232  -0.0036   0.0889
```

```
ATOM    5
 0.0232 -0.0064  0.0323 -0.0267 -0.4428  0.2134 -0.0177  0.1394 -0.0036  0.0889

ATOM    6
-0.0064  0.0326 -0.0001  0.0323  0.2134 -0.4049  0.1096 -0.0177  0.0486 -0.0073

ATOM    7
 0.0323 -0.0001  0.0326 -0.0064 -0.0177  0.1096 -0.4049  0.2134 -0.0073  0.0486

ATOM    8
-0.0267  0.0323 -0.0064  0.0232  0.1394 -0.0177  0.2134 -0.4428  0.0889 -0.0036

ATOM    9
 0.0889 -0.0073  0.0486 -0.0036 -0.0036  0.0486 -0.0073  0.0889 -0.3298  0.0765

ATOM   10
-0.0036  0.0486 -0.0073  0.0889  0.0889 -0.0073  0.0486 -0.0036  0.0765 -0.3298

END
```

B. Listings

```
      SUBROUTINE ATAT(N,M,ADIAG,C)
      DIMENSION ADIAG(30),C(30,30),PIRU(30,30)
      KLOW=M+1
      DO 12 IU=1,N
      DO 12 IR=1,IU
      S=0
      DO 10 J=1,M
      RJUJ=C(IR,J)*C(IU,J)
      DO 10 K=KLOW,N
      TERM=C(IR,K)*C(IU,K)/(ADIAG(J)-ADIAG(K))
   10 S=S+RJUJ*TERM
      PIRU(IR,IU)=S*4.
   12 PIRU(IU,IR)=S*4.
      WRITE(2,100)
  100 FORMAT(//27H ATOM-ATOM POLARIZABILITIES)
      DO 14 IU=1,N
      WRITE(2,101)IU
  101 FORMAT(/6H ATOM ,I4)
      WRITE(2,102)(PIRU(IR,IU),IR=1,N)
  102 FORMAT(10F8.4)
   14 CONTINUE
      RETURN
      END
```

3.5   PROBLEMS

1. Determine atom–atom polarizabilities for naphthalene using the NUCK matrix of Figure 3-3. Observe how the signs of $\pi_{r,u}$ alternate in passing from u through successive atoms r. Note that for comparable distances of separation the positive values, corresponding to u and r in opposite sets, exceed in absolute magnitude those appertaining to u and r in the same set. These properties can be associated through the relationship

$$\delta q_r = \pi_{r,u}\delta\alpha_u$$

with the 'law of alternating polarity', and, in the theory of chemical reactivity with *ortho-para* and *meta*-directing properties.

2. Apply the modifications $\delta\alpha_u = \pm 0{\cdot}125\beta$ to the naphthalene molecule with u in turn equal to 1, 2 and 9.

Establish that the polarizabilities $\pi_{r,u}$ are obtained to about four significant figures from the approximation formula

$$\pi_{r,u} \simeq \Delta q_r(\delta\alpha_u)/\delta\alpha_u$$

by comparing the results so obtained with those determined in problem 1.

Repeat the calculations with larger values of $\delta\alpha_u$, say $\pm 0{\cdot}250$, and establish that the accuracy in computing $\pi_{r,u}$ falls off.

Note that

$$\pi_{r,u}\,(+\delta\alpha_u) = \pi_{r,u}\,(-\delta\alpha_u)$$

for all r, u.

3. Obtain Hückel solutions for the data set given in Figure 3-3, and

(i) Confirm the 'pairing' properties listed in Section 3-1 for the AH obtained from the setting NMOD = 0.

(ii) Confirm the 'pairing' properties of conjugate solutions for related modifications $\pm\delta\alpha_u$ listed in Section 3-3, and verify that

$$\Delta q_s\,(-\delta\alpha_u) = -\Delta q_s\,(+\delta\alpha_u)$$

$$\Delta p_{st}\,(-\delta\alpha_u) = -\Delta p_{st}\,(+\delta\alpha_u) \qquad (s, t \text{ same set})$$

$$= +\Delta p_{st}\,(+\delta\alpha_u) \qquad (s, t \text{ opposite set})$$

for all atoms s and bonds s–t.

4. Compute charge densities, free valences, bond orders and atom atom polarizabilities for azulene

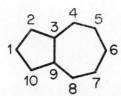

Verify that the $\pi_{r,u}$ do not alternate in sign in passing from u through successive atoms r.

5. Apply the modifications $\delta\alpha_u = \pm\beta$, $\pm 2\beta$ to a peripheral atom u of azulene, analogous to those applied to naphthalene in Figure 3-3. Compare the results obtained with those found for naphthalene, and verify that the properties characterizing 'conjugate' solutions are not found for azulene.

3.6 REFERENCES

1. C. A. Coulson and H. C. Longuet Higgins, *Proc. Roy. Soc.* (*London*), **A191**, 39 (1947).
2. C. A. Coulson and H. C. Longuet Higgins, *Proc. Roy. Soc.* (*London*), **A192**, 16 (1947).
3. H. H. Greenwood, *Brit. J. Cancer* **5**, 441 (1951).
4. H. H. Greenwood, *J. Chem. Phys.* **20**, 1333 (1952).
5. K. Fukui, T. Yonezawa and C. Nagata, *J. Chem. Phys.* **26**, 831 (1957).
6. C. A. Coulson and J. Jacobs, *J. Chem. Soc.* (*London*), 2805 (1949).
7. H. H. Greenwood and T. H. J. Hayward, *Mol. Phys.* **3**, 495 (1960).
8. C. A. Coulson and A. Streitwieser, *Dictionary of π-Electron Calculations* Pergamon, 1965.

4

Applications of Hückel Theory

Many aspects of experimental work on conjugated molecules can be interpreted by MO theory in its simplest, Hückel, form. Clearly complete agreement with experiment cannot be expected since the theory is based upon a simple model of the π-electron system alone, but many different types of applications demonstrate the effectiveness of the model, and frequently illuminate the nature and origin of physical and chemical properties. Typical examples have been widely described in scientific journals and textbooks. Streitwieser[1] has, for instance, provided an extensive and valuable review which discusses, in some detail, applications to the interpretation of dipole moments, bond lengths, nuclear quadrupole and electron-spin resonance spectra, oxidation–reduction and ionization potentials, chemical reactivity and electronic UV spectra. In most of these applications the objective is to find correlations between quantities defined in terms of a prescribed theoretical model and experimental results, and the computer programs described earlier can, in general, be used either directly, or with comparatively simple extensions, to calculate the relevant quantities.

A few trial runs will develop familiarity with the method of specifying conjugated molecules by means of the incidence matrix NUCK and the modification techniques, and, thereafter, no problems arise in the practical use of the programs. Some experimentation with 'standard' calculations in which groups $-NH_2$, $-OH$, $-NO_2$, $-COOH$, $-Cl$ etc., are attached to typical AHs, benzene, naphthalene, anthracene, can be recommended, since they provide useful evidence of times of computation for planning work systematically, and material for studying features characterizing Hückel solutions. It is important, of course, to recognize the appropriate 'parent' hydrocarbon which, when modified, yields the molecule whose

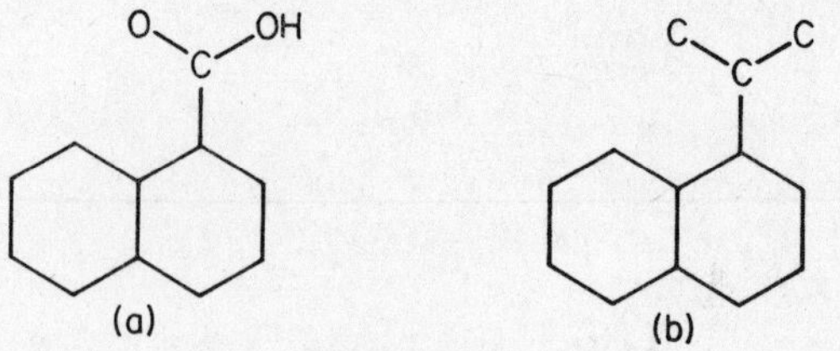

(a)   (b)

FIGURE 4-1

solution is required, as illustrated in Figure 4-1 by the $-COOH$ derivative of naphthalene.

In calculations of this kind coulomb integrals for atoms X other than carbon are expressed in the standard form

$$\alpha_X = \alpha + h_X\beta$$

and resonance integrals for bonds other than C–C, are written as

$$\beta_{X-Y} = k_{X-Y}\beta$$

where α and β are the carbon values. The parameters h_X and k_{X-Y} are introduced in the modification routine MODH as described earlier, and appropriate values may be obtained from journals and books, including Streitwieser's carefully and extensively documented review,[1] which contains the selected values given in Table 4-1.

Table 4-1

X	$\dot{N}$	$\ddot{N}$	N^+	$\dot{O}$	$\ddot{O}$	O^+	Cl
h_X	0·5	1·5	2·0	1·0	2·0	2·5	2·0

X – Y	C – $\dot{N}$	C – $\ddot{N}$	N – O	C – O	C = O	C – Cl
k_{X-Y}	0·8	1·0	0·7	0·8	1·0	0·4

A considerable effort has already been devoted to the derivation of appropriate parameters h_X and k_{X-Y} for heteroatoms and associated bonds by attaching various physical and conceptual notions to the interpretation of coulomb and resonance integrals. In the context of computer calculations, where results can be obtained rapidly, we prefer, in general, to compute sets of solutions for ranges of parameter values, and to select appropriate values retrospectively, by making comparisons between the numerical results obtained and experimental observations. Computed dipole moments, for example, restrict the ranges of plausible parameter values and can be used, in principle, for selection purposes, though comparisons with experimental results often raise problems (see Section 4-3). This kind of exploratory approach means that it is unnecessary to give interpretations to coulomb and resonance integrals which go beyond the definitions (2-16) and (2-17), though it is useful, in practice, to visualize these terms as being related in some way to electronegativities and interactions between adjacent atoms respectively; ultimately some sort of consistency amongst selected values must be found for the same atoms and bonds in different molecular environments. In adopting this approach we are not, thereby, rejecting the various interpretations that have, so far, been advanced in attempts to give values to coulomb and resonance integrals, but are simply shelving a lengthy discussion, comparable to that given in Streitwieser's excellent review,[1] in favour of parameter-scanning techniques.

It follows from these last remarks, that, even in the simplest applications, which aim to predict or explain experimental observations, computer methods of calculation are frequently used to generate sets of solutions of the Hückel equations as relevant parameters are varied, and such investigations usually develop into studies of the theoretical model itself in the context of the given problem. Therefore, since application of the computer programs to problems similar to those listed in the first paragraph of this chapter is generally straightforward, and MO interpretations based on computed results have been admirably described already,[1] the topics discussed in the rest of this chapter are chosen to illustrate the use of computers in studying properties of models and their bearing upon the ultimate theoretical interpretation. The choices may, at first sight, appear to be unrelated, and, from the point of view of the chemistry involved, this is certainly the case. Within the same chapter, for example, we discuss ideas relating to the localization of MOs (Section 4-2) and, elsewhere, the nature of conjugation in phosphonitrilic chlorides and similar molecules (Section 4-5). It is claimed that the former has some bearing upon the MO interpretation of many physical and chemical properties of conjugated molecules; and these ideas are, indeed, used in discussing $d_\pi-p_\pi$ conjugation in $(PN)_3Cl_6$. Similar techniques are also used in Section 4-1 in examining the nature of mesomeric effects in conjugated molecules R–Y, where R represents a parent hydrocarbon, and Y an attached group which conjugates with R. Thus, the theme which connects the different sections of this chapter is not, primarily, the physics and chemistry of the conjugated systems studied, but the design of experimental techniques for investigating theoretical models by computer methods.

4.1 MESOMERIC EFFECTS IN CONJUGATED MOLECULES

Many substituent groups with π-type $\phi(2p_z)$ orbitals in the valence shell conjugate with aromatic rings, and a valid formulation must include the group orbitals explicitly in the linear combinations

$$\psi = \sum c_r \phi_r$$

describing MOs. In consequence the dimensions of the secular equations and the numbers of π electrons change from those appertaining to the unsubstituted molecule, and new coulomb and resonance integrals for atoms in the substituent group may also be introduced. If styrene (Figure 4-2a) is regarded as a derivative of benzene both the dimensions of the secular equations and the number of π electrons changes from six to eight; in contrast, the dimensions in aniline (Figure 4-2b) augment to seven, and the number of π electrons to eight.

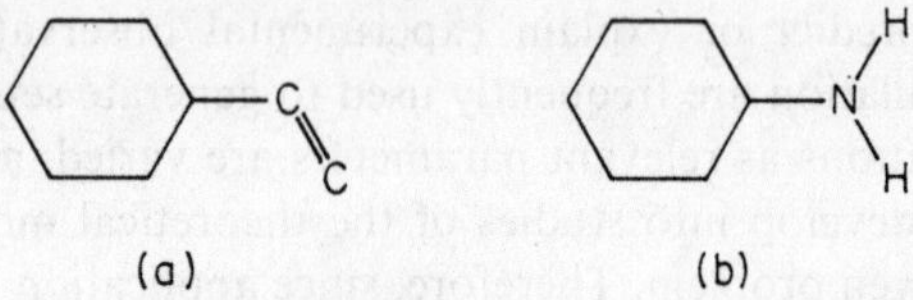

(a)   (b)

FIGURE 4-2

Consider the data specification for aniline with an incidence matrix NUCK describing the seven-membered bonding framework.

0001	NMOLS	
007004	N, M	

```
1  0
2  1 0
3  0 1 0
4  0 0 1 0           NUCK (80I1)
5  0 0 0 1 0
6  1 0 0 0 1 0
7  1 0 0 0 0 0
```

0003	NDER
0000	NMOD
0001	NMOD
0707 − 0·500	I, J, X
0000	NMOD
0707 − 1·000	I, J, X

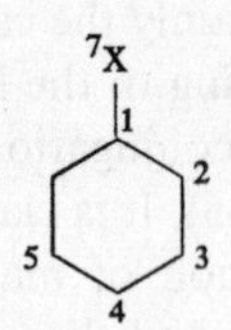

Here, aniline is referred to the 'benzyl' skeleton, and the sequence of modifications $\delta\alpha_7 = 0$, $0\cdot5\beta$ and $1\cdot0\beta$ generate firstly a solution for the benzyl negative ion, followed by two solutions for aniline. The bond order matrix is computed in subroutine PRS over the $M = 4$ lowest occupied orbitals.

Results for planar 1–3–5 triaminobenzene can be obtained by bordering the NUCK matrix for aniline, and adding modifications to appropriate diagonal elements

```
0001                    NMOLS
009006                  N, M
```

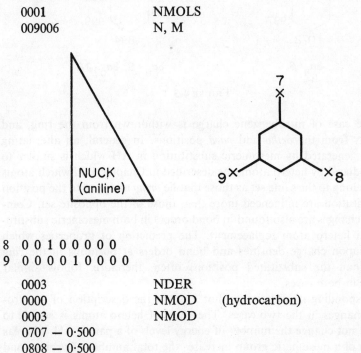

```
8  0 0 1 0 0 0 0 0
9  0 0 0 0 1 0 0 0 0
```

```
0003                    NDER
0000                    NMOD    (hydrocarbon)
0003                    NMOD
0707 — 0·500
0808 — 0·500
0909 — 0·500
        etc.
```

In this case $N = 9$, and the number $M = 6$ of doubly occupied orbitals accommodates the 12 π electrons.

Some interesting density-distribution patterns emerge from the numerical results obtained for substituted molecules, which have some bearing upon molecular properties. Consider the case of aniline, in which each carbon atom of the ring contributes one electron, and the nitrogen atom two electrons to the π-electron system. The nitrogen atom donates π-electron density to the ring, which appears as an increased charge density at the *ortho* and *para* positions, though, oddly enough, the density at the *meta* position is less than the value in benzene.

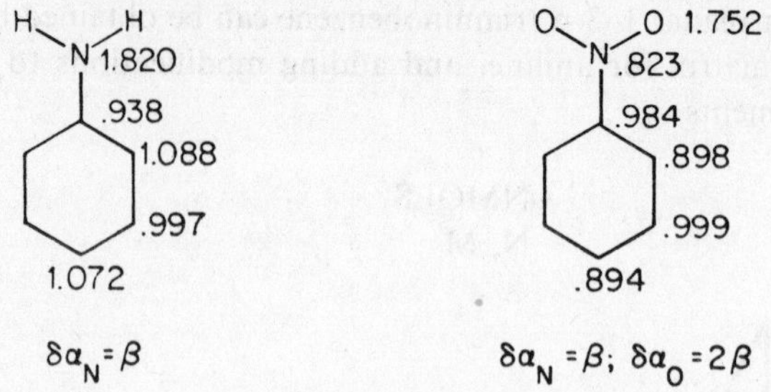

$$\delta\alpha_N = \beta \qquad\qquad \delta\alpha_N = \beta; \; \delta\alpha_O = 2\beta$$

FIGURE 4-3

In the case of nitrobenzene charge is withdrawn from the ring, and primarily from the *ortho* and *para* positions. In general, an alternating effect is generated by mesomeric substitution in AH which is similar to that produced by hetero atoms, as described in Chapter 3, in which atoms which belong to the same set as those that lie *ortho* and *para* to the position of substitution are influenced more than those of the opposite set. Comparable changes are also found in bond orders in both mesomeric substitution and hetero atom replacement. The prediction of properties which depend upon charge densities and bond orders at positions in the ring (other than the substituted position) often, therefore, follow similar patterns in both cases.

Corresponding similarities cannot arise in the description of spectroscopic changes in the two cases. The effect of hetero atoms is simply to shift but not change the number of energy levels of a parent AH, whereas addition of a mesomeric group increases the total number of orbitals and levels. The changes in energy-level diagrams can then be interpreted usefully in terms of a hypothetical 'fusion' of levels along the lines described in Chapter 2, especially with the aid of appropriate computer calculations. Such calculations have a direct bearing upon the interpretation of spectroscopic properties usually attributed to charge-transfer effects, and sets of computer solutions obtained as appropriate parameters are varied provide, as outlined below, a useful means of identifying 'charge-transfer' spectra.

The observed transitions of a conjugated molecule R–Y have sometimes been explained by the use of perturbation methods in terms of the MOs of the parent molecule R and of the mesomeric substituent Y. This approach has the merit of interpreting the observed spectra for the combined π-electron system R–Y by reference to that of R, so that the influence of the group Y may be recognized. In particular, it is often possible to classify transitions into 'local' excitations, that are largely confined to R (or Y), and 'charge transfer' excitations across R and Y. The MO bases of R and Y may not, however, be appropriate when π-electron delocalization between the two systems is appreciable, and characteristic properties of the spectra may, as a result, be obscured. The MOs of the combined system R–Y are then more suitable, and may be obtained from the standard computer programs for the Hückel method. It is still possible within this description to identify 'local' and 'charge-transfer' transitions when these arise, and to provide a more realistic interpretation than that given in terms of separate R and Y orbital bases. Indeed, since parameters can be varied easily and systematically within a single computer run, a simple set of calculations obtained by reducing in steps towards zero, the resonance integral of the bond joining R and Y will provide information on 'local' and 'charge-transfer' excitations, in more detail than can be provided by perturbation methods.

Several examples illustrating the properties of mesomeric substitution as discussed in this section are given in the form of problems at the end of the chapter. These and similar examples deserve careful study, since the computer solutions provide a wealth of information on how MO theory 'works' in practice, and particularly how it describes ground state and spectroscopic states of combined systems R–Y. The practical use of computer programs within this kind of context can convey many facets of the MO description, in particular, how the energy levels of combined systems R and Y interact, and how this interaction is reflected in the amplitudes of the corresponding MOs. It is highly desirable, if spectroscopic properties are to be investigated, to include the subroutine TRMOM described in Section 4-4. The results obtained from this program as the resonance integral between R and Y is varied systematically show how intensities of allowed transitions change as conjugation effects increase or diminish, and frequently provide interesting information on associated polarizations.

4.2 VARIATIONS OF α AND β

So far the effects of variations of parameters α and β have been studied in terms of perturbation theory, with special reference to the definition of polarizability coefficients, but this does not exhaust the available information on properties of the secular equations of the Hückel method. In particular certain, constraints operate that condition the nature of the solutions obtained when parameters are varied. These constraints are implicit, but are not specifically introduced in variations described by perturbation methods; as a result it is not possible to recognize, in such techniques, the influence of the constraints upon the form of solutions.

It is useful to introduce initially the idea of an energy band for AHs, within which all levels lie. Then, as coulomb integrals are varied, for example, all levels, with certain well-defined exceptions, are constrained to change towards limiting values which also lie within the band, and can be identified with energy levels of fragments of the original molecule, usually called 'residual' molecules. The variations considered in this section will be treated in a purely hypothetical fashion, though they relate directly to similar variations introduced, in perturbation and other methods, in describing molecular properties related to π-electron systems.

Consider firstly, the bounds for Hückel energy levels of AH. For linear polyenes $C_N H_{N+2}$ the levels are given by[2]

$$\epsilon_j = \alpha + m_j \beta$$

where

$$m_j = 2\cos[j\pi/(N+1)] \qquad (j = 1, 2, 3 \ldots N)$$

with bounds $-2 \leqslant m_j \leqslant +2$. As N becomes large, the levels can be visualized as forming a band within these bounds. For cyclic polyenes, the corresponding formula for the levels[2]

$$m_j = 2\cos(2\pi j/N) \qquad \left(j = 0, \pm 1, \ldots \pm \left[\frac{N}{2} - 1\right], \frac{N}{2}\right)$$

yields similar bounds for the band, and for the linear polyacenes[3]

$$m_j = 1, \tfrac{1}{2}[1 + \sqrt{9 + 8\cos(j\pi/(N+1))}\,]$$

$$-1, -\tfrac{1}{2}[1 + \sqrt{9 + 8\cos(j\pi/(N+1))}\,]$$

with bounds $-\tfrac{1}{2}(1 + \sqrt{17}) \leqslant m_j \leqslant \tfrac{1}{2}(1 + \sqrt{17})$ or, approximately $-2 \cdot 5 \leqslant m_j \leqslant 2 \cdot 5$. It can be shown that, for graphite[4] $-3 \leqslant m_j \leqslant 3$, and these may be interpreted as general bounds for AH.

A. Changes of coulomb integral

In an AH all coulomb integrals are equal to α the chosen zero of energy, and all resonance integrals β are the same; the energy levels are distributed symmetrically about α (Figure 4-4), and the charge density at each conjugated atom is unity. Consider the effect of changing the coulomb integral α_r at the rth position, where (equations 2-44, 45)

$$\alpha_r = \alpha + \delta\alpha_r$$

and

$$\delta\alpha_r = h_r\beta$$

Any atom r of an even AH defines a corresponding 'residual molecule' RM_r as the odd AH obtained by excluding atom r from the original conjugated system. The secular determinant of the RM_r can be obtained by crossing out the rth row and column of the determinant $\Delta(\epsilon)$ of the original system (equation 2-19). Denote the reduced determinant by $\Delta_{r,r}(\epsilon)$ in agreement with the notation used earlier; then the energy levels of the RM_r are obtained from the roots of

$$\Delta_{r,r}(\epsilon) = 0$$

and the MOs by solving a corresponding set of secular equations for each root. It can be proved that, for all possible RM_r corresponding to all conjugated atoms r of the system, the energy levels 'separate' those of the parent AH as shown in Figure 4-4; the positioning of the levels depends upon which atom r is chosen. A central level always lies in the zero of energy α, and the corresponding MO is usually termed the 'non-bonding orbital' NBMO (Chapter 3, Section 3-1), In fact, the polynomial $\Delta_{r,r}(\epsilon)$ contains odd powers of ϵ only, giving a root $\epsilon = 0$, and the structures of energy levels can be readily understood, at least intuitively, in terms of related ideas discussed already in Chapter 2.

The same set of energy levels for the residual molecule RM_r appear on both the left and right-hand sides of Figure 4-4, and relate to variations of $\delta\alpha_r$ of opposite signs, as described below. The line AA' lies at 45° to the horizontal axis on which $\delta\alpha_r$ is measured, provided the ϵ and $\delta\alpha_r$ scales, both of which are measured in units of β, are the same.

Suppose now that h_r is increased positively from the value zero, so that α_r becomes increasingly negative. Then the following properties characterize solutions of the secular equations.[5,8]

(1) All energy levels are lowered. The lowering of the deepest level ϵ_1

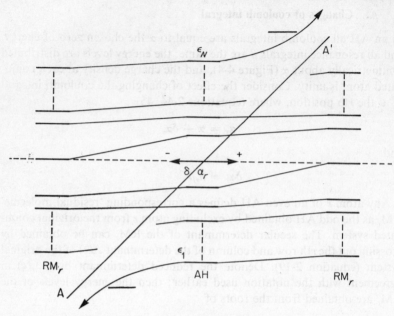

FIGURE 4-4

is limited only by the magnitude of $\delta\alpha_r$, and tends to the 'asymptote' AA' as $\delta\alpha_r \to -\infty$. The remaining levels are lowered towards those of the RM_r as limiting values, as shown in the left side of Figure 4-4.

(2) The MOs change so that ψ_1 which corresponds to the lowest energy level ϵ_1 becomes increasingly localized at atom r, and the amplitudes of all other MOs at atom r are reduced; in fact, all other MOs tend increasingly towards those describing the RM_r. These conditions are achieved by shifts in the positions of all the nodes of the MOs; a node of each MO 'migrates' until it cuts the bonds connecting atom r to the RM_r. In a sense, therefore, the number of nodes embraced by the RM_r is, in the limit, reduced by one, and the correct nodal character over the RM_r is thereby preserved.

(3) The charge density q_r at atom r increases; this implies that the increase in density of the lowest orbital ψ_1 at atom r exceeds the decrease in density at the same atom, of the remaining orbitals. The limiting density $q_r = 2$ is obtained for $\delta\alpha_r = -\infty$.

The effects described in (2) are sketched in Figures 4-5(a) and (b) for the three lowest levels of a linear chain of N-conjugated atoms, in which, for simplicity, the modified atom $r = N$ is taken as a terminal atom. The horizontal axis identifies the positions of atoms, labelled $1, 2, \ldots N$, within the molecule. Since the potential, though non-uniform in Hückel theory, assumes a similar form around each carbon atom, the same axis may be used, purely symbolically, to identify modified regions. Thus a lowering $\delta\alpha_r$ in coulomb integral at the terminal atom $r = N$ is recognized by the 'well' $\delta\alpha_N$ sketched in Figure 4-5(b).

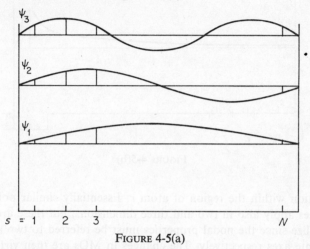

FIGURE 4-5(a)

The remaining horizontal lines provide axes for the diagrammatic representation of the MOs. If the amplitudes c_{sj} are plotted as vertical line segments at each atom position, as shown for a few atoms s, then the MOs ψ_j are represented by the envelopes. These diagrams are not drawn to scale, nor should they be identified with similar diagrams used in free-electron theories. They are simply representations which depict the changes in MOs described above, particularly the tendency towards 'localization' of a single orbital within the region $r = N$, and the simultaneous preservation of the correct nodal character of each orbital over the residual molecule.

The modifications of MOs of a linear chain due to changes in coulomb integral of a terminal atom can be sketched easily. It is interesting, however, to pursue the problem further by producing similar diagrams when the modified atom r lies within the body of the molecule. Computer calculations based on a NUCK matrix for a linear chain of say $N = 10$ atoms, in which the coulomb integral of the rth is modified in the subroutine MODH in large steps, $\delta\alpha_r = 2\beta, 4\beta$ and 6β say, will provide a clear account of the emergence from the band of an 'impurity' level, and of orbital

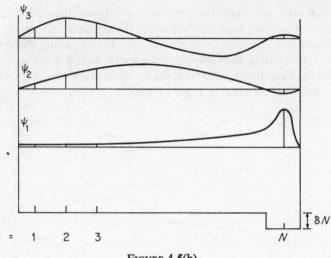

FIGURE 4-5(b)

localization within the region of atom r. Essentially similar polarization properties apply also in two and three dimensions, but these are difficult to visualize since the nodal properties must be referred to two and three coordinate axes respectively. The changes in MOs are then virtually impossible to sketch, but can be checked and recognized when appropriate computer calculations have been made.

We now return to consider changes in the energy-level diagram which lie to the right side of Figure 4-4, and are obtained by making h_r negative, and hence $\delta\alpha_r$ positive, a change of coulomb integral opposite in sign to that producing the left side of the diagram. In fact, the diagram is symmetrical about the origin, $\delta\alpha_r = 0$, in the sense that antibonding levels on one side are mirror images in the zero of energy, of bonding levels on the other side—a result already established in Chapter 3, and described there in terms of 'conjugate' solutions. Clearly, the uppermost level ϵ_N emerges from the 'band', and is restricted only by the value of $\delta\alpha_r$, while the remaining levels change towards those of the same residual molecule RM$_r$. The MO ψ_N which has $(N - 1)$ nodes now becomes increasingly localized in the region of atom r, and the remaining MOs change towards those of the residual molecule.

Variations of a coulomb integral are therefore associated with a progressive localization of a MO in the region of the modified atom r and a simultaneous decrease in amplitude at the same position of all other MOs, and this process of localization is ultimately accompanied by the emergence of the corresponding energy level from the 'band'. These ideas obviously extend to the modification of two or more coulomb integrals which bring about partial localization of two or more orbitals at modified atoms, and the emergence of the same number of energy levels from the band.

These properties find their origin, of course, in the MO formulation. The orthogonality or independence of ψs makes it possible to describe and recognize the influence of the Pauli exclusion principle in controlling the π-electron distribution; it means that polarization processes occur physically so that not more than two electrons, with opposed spins, may occupy the same region of space.

B. Changes of resonance integrals

The idea of localization of orbitals can be pursued further, in the same purely exploratory way, as properties of the secular equations, by varying other parameters.

Suppose, firstly, that the resonance integrals associated with a conjugated perimeter atom r and its two neighbours s and t are reduced from the standard carbon value β according to the formula

$$\beta_{r\sigma} = k\beta \qquad (\sigma = s, t)$$

and $k = 1 \to 0$. The energy levels 'contract' from those of the parent AH towards those of the RM_r defined by excluding atom r as shown in Figure 4-6.

The innermost pair of levels produce, in the limit when $k = 0$, a doubly degenerate level with energy $\epsilon = \alpha = 0$. The set of MOs change, as $k \to 0$, towards those of the RM_r except the two ψs corresponding to the innermost pair of levels, which retain a uniform distribution over the complete parent hydrocarbon. Alternative descriptions may be obtained, however, by taking orthogonal combinations of this pair of MOs when $k = 0$; in particular, sum and difference combinations yield the NBMO of the RM_r and a π orbital confined to atom r. It is, therefore, legitimate to visualize some form of 'localization' at the centre of the band associated with the reduction of resonance integrals between a conjugated atom r and its neighbours.

The changes in energy levels have been described above without reference to their occupancies by π electrons. When $k > 0$ the bonding levels that lie below the zero of energy α are clearly doubly occupied in the ground state; but for $k = 0$ the allocation of π electrons to the least

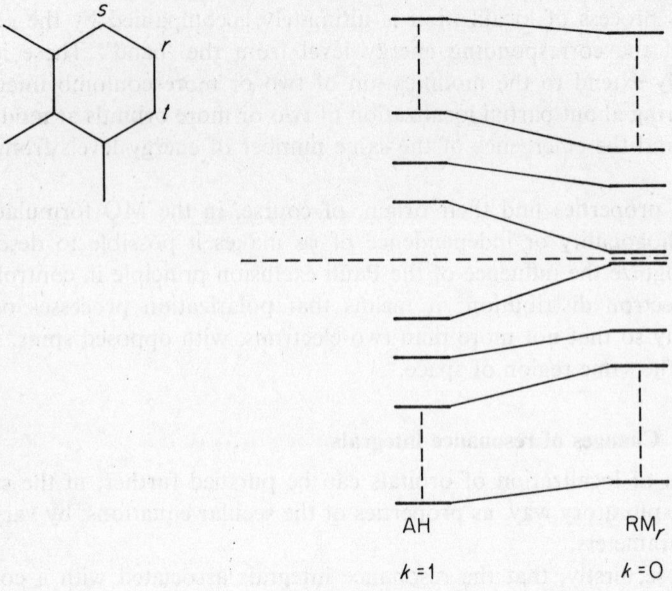

FIGURE 4-6

bound level becomes ambiguous, since two levels are available for two π electrons, and more than one possible π-electron configuration with resultant spin zero, possesses the same total π-electron energy. In such situations Hückel theory breaks down, and the problem can be resolved only by invoking CI methods. It is, of course, unrealistic to assume that Hückel theory fails only for exact degeneracies; the method must break down for near degeneracies also, and situations of this kind arise not infrequently in theoretical models employing the method.

C. Bond 'localization'

Consider now a change in resonance integral β_{rs} of a perimeter bond joining atoms r and s given by

$$\beta_{rs} = k\beta$$

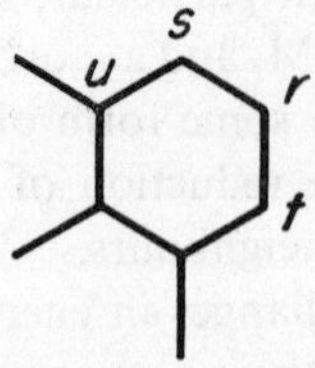

where k is assumed to increase from the value 1 in the parent to larger values. Two levels eventually emerge from the band, one above and one below. The MOs associated with these levels increasingly acquire the forms

$$\psi_{\text{lower}} \propto [c_r\phi_r + c_s\phi_s + \sum_{\lambda \neq r,s} c_\lambda\phi_\lambda]$$

$$\psi_{\text{upper}} \propto [c'_r\phi_r + c'_s\phi_s + \sum_{\lambda \neq r,s} c'_\lambda\phi_\lambda]$$

where
$$c_r \simeq c_s \gg |c_\lambda|$$

$$c'_r \simeq -c'_s \gg |c'_\lambda| \qquad \text{(taking } c'_r \text{ positive)}$$

The results obtained imply that the emerging levels become increasingly analogous to the levels of a 'localized' diatomic π bond associated with atoms r and s, and are strongly bonding and antibonding due to the large value of β_{rs}; indeed, when normalized, the coefficients c_r and c_s (and c'_r and $-c'_s$) tend towards the diatomic value of $1/\sqrt{2}$.

Another way in which a 'diatomic' π bond r–s can be localized is by reducing the resonance integrals β_{rt} and β_{su} towards zero. In the limit, when $\beta_{su} = \beta_{rt} = 0$ two energy levels associated with a diatomic π bond connecting atoms r and s take the values $\pm\beta$ as in ethylene, and $c_r = c_s = c'_r = -c'_s = 1/\sqrt{2}$. It is possible, in both forms of bond localization, to regard the π-electron fragment obtained by excluding atoms r and s as the appropriate residual molecule, and confirmation of these descriptions should be sought by carrying out sets of computer calculations with appropriate variations of bond parameters.

D. Some areas of application

The various 'processes' of localization discussed above, though at first sight hypothetical in character, are frequently implicit in descriptions of chemical and physical properties of π-electron systems. Indeed many applications of the MO method refer to changes in π-electron configurations which are made under constraints similar to those outlined above.

Take, for example, the simple case of replacement of a carbon atom by a heteroatom such as nitrogen in pyridine, quinolene, and similar molecules. The replacement results in a shift of π-electron charge towards the N atom, which is, however, composed of an increase in density at the N atom of the π-electron MO of lowest energy, and a decrease in density at the same atom of all other MOs. Clearly, the increase of the lowest orbital must exceed the total decrease in density at the N atom of all other orbitals to ensure a net flow of charge towards the nitrogen atom. A similar change in π-electron distribution occurs in the MO description of the polarization produced by approaching ionic reagents. According to the isolated molecule method the π-electron charge shifts produced by a neighbouring charged reagent brings to the position under attack an increased density of one MO, depending upon the sign of the charge carried, and the densities of all other MOs at the same position are simultaneously reduced (Chapter 5).

Resonance integrals may be assumed to change when rotation about conjugated bonds occurs. Take, for example, the case of stilbene and assume that large groups (e.g. CH_3) attached to positions 3 and 5 do not

conjugate with the ring, but prevent planarity, which is relieved by rotation about the bond 4–13. The axes of atomic orbitals $\phi_4(2p_z)$ and $\phi_{13}(2p_z)$ are then no longer parallel, and rotation can, therefore, be represented as a reduction of the resonance integral

$$\beta_{4-13} = \int\phi_4(2p_z)h_\pi\phi_{13}(2p_z)\,d\tau$$

or

$$\beta_{4-13} = \beta \cos \theta$$

where β is the standard carbon resonance integral, and θ the angle between the axes of the $\phi(2p_z)$ atomic orbitals on atoms 4 and 13. It is useful to examine by computer calculations changes in π-electron configuration as β_{4-13} is reduced systematically in steps towards zero; in the limiting case the solution describes two conjugated fragments which correspond to benzene and styrene π-electron systems. It is interesting to note that throughout these changes the charge densities q_r at each atom r remain unity, whilst individual orbitals and levels change towards those of the fragments. We can operate the 'process' in reverse and easily deduce, from the 'repulsions' amongst energy levels that the lowest excitation energy in stilbene is smaller than that of both benzene and styrene. Thus rotation about a 'single' π bond in stilbene is associated with a shift towards the blue end of the spectrum.

Although planarity is usually relieved by rotation about 'single' bonds which are weaker than partial double bonds, it is interesting, nonetheless, to examine from a theoretical point of view what happens when β_{13-14} is reduced systematically towards zero. This problem is left as an exercise (problem 8 at the end of the chapter) for the reader; here we note simply that, in the limit as $\beta_{13-14} \to 0$ the π-electron configuration must describe two 'benzyl' π-electron systems. Each benzyl system has an energy level lying in the zero of energy and, therefore, rotation about a partial double bond, in stilbene, for example, is associated with a shift to the red end of the spectrum.

Mulliken[6] has given descriptions of hyperconjugation in the transition state of electrophilic reactions by drawing upon the equivalence between

$$\text{(A)} \qquad or \qquad \text{(B)}$$

the σ-bond complex (A) and the pseudo π-bonded system (B), as described later (Chapter 5). Two electrons are withdrawn from the benzene system to form a σ-type bond with X, the incoming electrophilic reagent, and therefore, in the equivalent π-electron model, the transition-state configuration is described in terms of four π electrons associated with the pentadienyl residual molecule, and two with the pseudo π bond. Hyperconjugation is then described in terms of a limited delocalization effect operating between the two π-electron fragments, and is represented formally by introducing small resonance integrals $\beta_{rt} = \beta_{ru}$ between atom r and its neighbours t and u in the residual molecule. Mulliken assigned a large resonance integral to the pseudo π bond $r-s$, to retain within $r-s$ a high degree of double bond character when hyperconjugation is introduced, and the theoretical model is, therefore, closely related to the form of bond localization described in Section C.

Brown's[7] definition of bond localization energy which has been used as a reactivity index (Chapter 5) for describing the reactions taking place at conjugated bonds, is also based on a model which is closely related to the description of bond localization given in Section C. It is, however, characteristic of theoretical interpretations of reactivity in conjugated molecules, as presented hitherto, that, although modifications in π-electron distributions are assumed, associated changes in electron energy levels and orbitals are seldom described. Ultimately, some form of localization, perhaps partial localization, of orbitals must, however, develop to prepare for spin-coupled electron-pair bonding with the incoming reagent at the position of attack. Within this contect, therefore, the emergence of levels from the 'band' and the associated localization of orbitals as described above may enhance, and possibly provide essential features of physically acceptable theoretical models.

An important area of application which can be mentioned here only briefly, concerns the MO theory of surface states.[9,10,11] The potential at surface atoms differs from that within the body of the solid, and surface states are associated with levels that, as a result, emerge from the energy band. The levels that emerge and their associated orbitals can be enumerated and identified in relation to the complete set of orbitals and levels for the lattice by extending the qualitative description given earlier for a one-dimensional model. Consider, for example, the case of a two-dimensional rectangular lattice with sides containing M and N atom sites. The MOs ψ_{mn} for the lattice can be classified by the subscripts which identify the number of nodes referred to rectangular areas which lie parallel to the lattice planes, where $m = 0, 1, 2, \ldots (M - 1)$ and $n = 0, 1, 2, \ldots (N - 1)$. Suppose that, initially, the same potential applies at all lattice sites; then assume that the potential is lowered equally at all atoms in one side containing N atoms. The N energy levels associated with the sequence of orbitals

$$\psi_{00}, \psi_{01}, \psi_{02}, \psi_{03}, \cdots \psi_{0N-1}$$

emerge successively from below the band as the surface potential is lowered. Each orbital ψ_{0n} is nodeless in a direction perpendicular to the modified edge with an amplitude variation, in this direction, similar to that depicted for a modified end atom of a linear chain, by ψ_1 in Figure 4-5(b). Thus the number of surface states which are associated with emerging levels and partial orbital localization at surface atoms, is identical to the number of modified surface atoms. These ideas are readily extended to modifications of more than one surface in two and three-dimensional lattices. Ultimately, however, the magnitude of modifications in surface potentials which produce surface states which are associated with energy levels that emerge from the band, can only be determined by a more complete theoretical investigation. The purpose of these brief comments is simply to show that useful information concerning the enumeration and origin of surface states can be obtained by adapting and extending the idea of orbital localization as described earlier.

In this section we have indicated the importance in certain areas of application of calculating the patterns of energy-level and orbital changes implicit in theoretical models when parameters vary, and when the existence of computing facilities enables this to be done rapidly and extensively. It would be no exaggeration to suggest that this procedure should be followed for all π-electron MO problems, even for those that can be solved approximately and rapidly by perturbation methods, since the complete solutions provide so much more information.

4.3 DIPOLE MOMENTS IN CONJUGATED MOLECULES

The main purpose of this section is to outline the theoretical basis on which the π-electron dipole moment of a conjugated molecule is calculated in the subroutine DIMO listed at the end of the chapter. However, the π-electron moment, when computed automatically by the program, should be used with discretion, not merely because Hückel theory appears to exaggerate, in certain circumstances, inequalities in the charge distribution, as indicated in the case of azulene discussed later, but also because, in making comparisons with experiment, real problems arise in separating the observed moment into π and σ components. The procedure for finding components is often more difficult than generally supposed, especially when hetero atoms with lone-pair electrons participate in conjugation, as indicated later.

The contribution to the total dipole moment of a conjugated molecule of the π-electron system in the ground-state configuration Ψ_0 is given by

$$\mu_\pi = e \int \Psi_0^*[-\sum_s Z_s R_s + \sum_i r_i]\Psi_0 \, d\tau \qquad (4\text{-}1)$$

where R_s is the position vector of the sth conjugated atom and r_i that of the ith π electron. Z_s is the effective nuclear charge at atom s and is unity in a neutral atom contributing one π electron to the system. The term

$$\bar{r} = \int \Psi_0^*(\sum_i r_i)\Psi_0 \, d\tau \qquad (4\text{-}2)$$

represents the centre of gravity of the π-electron distribution in the ground-state configuration, and has components parallel to the coordinate axes, for example

$$\bar{x} = \int \Psi_0^*(\sum_i x_i)\Psi_0 \, d\tau \qquad (4\text{-}3)$$

with similar expressions for $\bar{y}$ and $\bar{z}$ such that

$$\bar{r} = \sqrt{\bar{x}^2 + \bar{y}^2 + \bar{z}_2} \qquad (4\text{-}4)$$

The first term in the bracket (4-1) can be taken outside the integral, and represents the centre of gravity of the effective charges. It can be shown that, when the ground-state wavefunction is introduced, in either the product or determinantal form, the expression (4-3) reduces, because of orthogonality amongst the MOs ψ, to

$$\bar{x} = \sum_i \int \psi_i x \psi_i \, d\tau \qquad (4\text{-}5)$$

where the summation runs over all doubly occupied orbitals. Expanding each ψ in terms of atomic orbitals $\phi_s(2p_z)$

$$\psi_i = \sum_{s=1}^{N} c_{si}\phi_s \qquad (4\text{-}6)$$

gives

$$\bar{x} = \sum_{s=1}^{N} q_s R_s^x \qquad (4\text{-}7)$$

since the cross terms

$$\int \phi_s x \phi_t \, d\tau \qquad (4\text{-}8)$$

are assumed to be zero, in accordance with the overlap approximation introduced in Hückel theory. In equation (4-7) R_s^x is the x component of R_s, the position vector of the sth atom, and q_s is the associated π-electron charge density.

For a neutral atom s contributing one π electron to the conjugated system, $Z_s = 1$, and

$$\mu_\pi^x = e\sum_s (q_s - 1)R_s^x \qquad (4\text{-}9)$$

is the x component of the total moment μ_π, calculated as the sum of products of net charge $(q_s - 1)$ and the x coordinate at each atom.

Similar expressions determine μ_π^y and μ_π^z, which is zero for a planar molecule, and hence

$$\mu_\pi = \sqrt{(\mu_\pi^x)^2 + (\mu_\pi^y)^2 + (\mu_\pi^z)^2} \qquad (4\text{-}10)$$

The subroutine DIMO listed at the end of this chapter, calculates the dipole moment for planar-conjugated molecules from the expression

$$\mu_\pi = \sqrt{(\mu_\pi^x)^2 + (\mu_\pi^y)^2} \qquad (4\text{-}11)$$

It can be called in the usual way by the Hückel main program as indicated later, provided atom coordinates are available. Provision is also made for inclusion of Z_s values other than unity, which apply when an atom contributes zero or two π electrons.

Carbon atoms in AH are computed, in the Hückel method, to be electrically neutral, since the effective unit positive charge at each nucleus in the σ-bonded framework is neutralized by a corresponding π-electron density of unity. Computed values of charge densities in non-alternants, heterocyclics, and substituted molecules are not unity, and net dipole moments, due to asymmetries in π-electron charge distributions, result. These π-electron dipole moments should, in principle, provide either a useful test of the theory when compared with experimental results, or a means of deducing appropriate parameters, according to the context. The model works quite well in practice, though certain applications present rather stringent tests, and the computed results can then be far from satisfactory. For example, for azulene, Hückel theory computes a π-electron moment of about six Debye Units, and no obvious source of σ-bond moment exists that can result in a total moment of about 1D to compare with that observed experimentally. Hückel theory fails to prevent the

Charge densities in azulene; $\mu_\pi \sim$ 6D.

build-up of excessive charge densities in various regions of the molecule, by neglecting interelectron repulsions amongst π electrons, and SCF theory (Chapter 6), which includes these repulsions explicitly, demonstrates the magnitude of the error by reducing the calculated π-electron moment to less than 2D. However, if the resonance integral, β_{3-9} corresponding to the long bond bridging the molecule, is reduced from the standard carbon–carbon value β, in the Hückel description, the calculated dipole moment diminishes correspondingly. The effect can readily be recognized, since, as this particular resonance integral is reduced towards zero, the 'molecule' is progressively transformed into an AH with unit charge density at each atom and zero dipole moment. Thus deficiencies in the calculation of dipole moments by the Hückel approximation are mainly inherent in the method itself, but may also be parameter-dependent.

However, additional difficulties always arise when comparisons are made with experimentally determined dipole moments, since the observed moment must be decomposed into σ and π contributions. Consider the case of pyridine (Figure 4-7) where the σ-bond framework is formed by

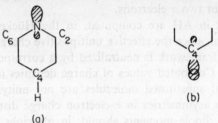

(a)

(b)

FIGURE 4-7

sp^2 trigonal hybrid atomic orbitals, with axes lying in the molecular plane at 120° separation. The hybrid orbitals of carbon atoms external to the hexagonal ring form C–H bonds, and the corresponding nitrogen hybrid accommodates the 'lone-pair' electrons. In making comparisons with calculated π-electron moments it is necessary to separate σ and π contributions to the total observed dipole moment of 2·21 Debye Units. The σ-moment contribution originates largely as the resultant moment of the lone-pair electrons on the nitrogen atom and of the C_4–H bond opposite, since the remaining σ-bond moments cancel, or are comparatively small. The moment due to the lone pair lobe and the corresponding two positive charges located on the N nucleus is around 3·5 D, and that of a trigonal C_4–H non-polar bond is about 2·0 D. In both cases these are computed from the spatial distribution of two electrons and the centres of gravity of the two corresponding positive charges; the large C–H bond moment is, for example, attributable to the highly asymmetric trigonal lobe of the C-bonding orbital (Figure 4-7b). Thus the maximum value of μ_π in pyridine is of the order

$$\mu_\pi \simeq 2\cdot2 - [3\cdot5 - 2\cdot0] = 0\cdot7 \text{ D}$$

In fact C–N σ-bond moments (C $\equiv$ C$_2$, C$_6$) could reduce this value by an unknown amount, and the percentage uncertainty in μ_π is, therefore, comparatively large.

When N substitution is represented by the modified coulomb integral

$$\alpha_N = \alpha + \delta\alpha_N$$

with

$$\delta\alpha_N = h_N\beta$$

it is found that μ_π is effectively linear with respect to variation of h_N over the range $0 \leqslant h_N \leqslant 0\cdot5$; then μ_π can be written in a form, approximate to the first order in $\delta \alpha_N$

$$\mu_\pi \text{ [pyridine]} = 2 \cdot 216 \, h_N \qquad \text{Debye Units.}$$

which covers the region of the expected value. The constant $2 \cdot 216$ is derived from the dimensions and geometry of pyridine, assuming equal bond lengths of $1 \cdot 40$ Å. When $h_N = 1$, μ_π (exact) $= 2 \cdot 083$ D which shows that higher order terms are small even for this comparatively large value of h_N.

It may be helpful to note, at this point, that traditional vector methods for combining bond moments do not deal explicitly with lone-pair moments, and the contributions of bonds defined by the distribution of electron pairs forming bonds and their corresponding positive charges.[12] In ammonia, for example, the observed moment of $1 \cdot 5$D is resolved into components directed along equivalent N–H bonds, which are then assumed

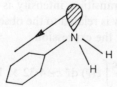

(a)   (b)

to be ionic (a), with polarity N^-H^+. In fact, the observed moment can be accounted for almost entirely in terms of the spatial distribution of lone-pair electrons ($\mu_{lp} \sim 3 \cdot 5$D) and covalent electron-pair bonds ($\mu_{\text{bond}} \sim$ 2D), which lie (b) roughly along tetrahedral directions. The traditional method lumps together lone pair and electron-pair bond moments in deriving N–H vector components which are assumed to arise entirely from ionic bonds, and, though the origin of bond moments in this interpretation is generally unacceptable, vector addition of components derived in this way appears to produce satisfactory values for resultant dipole moments. However, in molecules like aniline where lone-pair electrons participate to some extent in conjugation, the validity of the method is less obvious. Aniline is not planar, though the properties of the molecule are largely conditioned by delocalization of the lone-pair electrons over the adjacent ring, and, therefore, subdivision of the observed dipole moment into component parts requires careful consideration of the principles invoked.

Thus although traditional vector methods apparently work satisfactorily in practice, it may be important in certain applications to examine the origin of dipole moments more closely in terms of charge asymmetries, and charge shifts, such as occur in π-electron systems and ionic bonds. Problems can certainly arise in subdividing observed moments into σ and π component parts that allow comparison with calculated estimates of π-electron dipole moments.

4.4 TRANSITION MOMENTS IN π-ELECTRON SYSTEMS

The two main features characterizing π-electron absorption spectra are the frequencies and intensities of absorption, and these are related, since the intensity is defined (equation 4-16) theoretically in terms of the frequency.

Consider the 'excitation' process in which a π electron is transferred from an occupied orbital of the ground-state configuration to an unoccupied, or virtual, orbital to produce an 'excited' π-electron configuration. In Hückel theory the energy of excitation $\Delta\epsilon$ from the ground state is simply the difference in energy of the levels of the two orbitals concerned,

$$\Delta\epsilon(i \to k') = \epsilon_{k'} - \epsilon_i \qquad (4\text{-}12)$$

where i is the orbital doubly occupied in the ground state, and k' the virtual orbital, denoted by a prime.

The corresponding frequency $\nu_{i \to k'}$ of absorption is given by

$$\Delta\epsilon(i \to k') = \mathbf{h}\nu_{i \to k'} \qquad (4\text{-}13)$$

or, in terms of wavenumber $\bar{\nu}$ (cm^{-1}) and wavelength λ by

$$\Delta\epsilon(i \to k') = \mathbf{hc}\bar{\nu}_{i \to k} = \mathbf{hc}/\lambda_{i \to k}' \qquad (4\text{-}14)$$

where $\mathbf{c}$ is the velocity of light, and $\mathbf{h}$ is Planck's constant. The quantity used to characterize the transition intensity is the oscillator strength f. This dimensionless quantity is related to the observed intensity distribution in the absorption band, by the expression

$$f_{\text{obs}} = \frac{2 \cdot 303\mathbf{c}}{Ne^2} \int \xi(\bar{\nu}) \, d\bar{\nu} \simeq 4 \cdot 32 \times 10^{-9} \int \xi(\bar{\nu}) \, d\bar{\nu} \qquad (4\text{-}15)$$

where $\xi(\bar{\nu})$ is the molar extinction coefficient, measured at wavenumber $\bar{\nu}$. An expression for the oscillator strength can be derived theoretically in the form

$$f_{\text{theor.}} = \left(\frac{8\pi^2 mc}{3h}\right)\bar{\nu}Q^2 = 1\cdot085 \times 10^{11}\bar{\nu}Q^2 \qquad (4\text{-}16)$$

where

$$Q = \int \Psi_{\text{exc}}^*(\sum_i r_i)\Psi_0 \, d\tau \qquad (4\text{-}17)$$

is the transition moment connecting the ground-state configuration Ψ_0 and the excited configuration Ψ_{exc}. The summation is taken over the position vectors r_i of all the electrons concerned, in this case, the π electrons. Q is a vector quantity with components Q_x, Q_y and Q_z given by

$$Q_x = \int \Psi_{\text{exc}}^*(\sum x_i)\Psi_0 \, d\tau \qquad (4\text{-}18)$$

with similar expressions for Q_y and Q_z, and

$$Q = \sqrt{Q_x^2 + Q_y^2 + Q_z^2} \qquad (4\text{-}19)$$

If the transition is G-fold degenerate, then the expression (4-16) is generally multiplied by G on the right-hand side.

The excited π-electron configuration $\Psi(i \to k')$ differs from that of the ground state Ψ_0 by replacement of the occupied orbital ψ_i by the virtual orbital $\psi_{k'}$ as implied by the energy of excitation (4-12). It can be shown that in this case the expression (4-18) reduces, because of orthogonality amongst the MOs ψ, to

$$Q_x = \sqrt{2}\int \psi_{k'} x \psi_i \, d\tau = \sqrt{2}m_{ik'}^x, \qquad (4\text{-}20)$$

and similar expressions are obtained for Q_y and Q_z. It may be noted that whereas $\bar{x}$ in (4-5) represents an average value taken over all doubly occupied π-electron MOs, as expected, a similar reduction from (4-18) gives Q_x as an average value between two MOs only, namely the two orbitals involved in the excitation 'process'.

The transition moment Q is zero when all its components are zero, and this condition defines a so-called 'forbidden' transition, of zero intensity. Such conditions are the origin of selection rules for spatial parts of the wavefunction. In molecules with geometric symmetry, the MOs ψ_i and $\psi_{k'}$ which determine components of Q, themselves belong to some symmetry species, and selection rules for transitions can be determined on inspection by group theoretical rules. However, in the context of computing these techniques are seldom of value. It appears that, in general, each symmetry type must be programmed individually for application to appropriate molecules, which, more seriously, cannot be modified, since the assumed symmetry is, generally, thereby destroyed. Group theoretical methods are, in fact, of doubtful value in computing, and it is usually simpler and preferable to write general routines which evaluate oscillator strengths for all transitions specified, including those that turn out to be forbidden.

We now turn to a consideration of the use of the subroutine TRMOM for calculating intensities and polarizations of π-electron transitions in conjugated molecules within Hückel theory. It is well known that the Hückel method fails to provide a theoretically valid calculation of energies of excitation in AH, but since transition moments are easily calculated from the eigensolution of the secular equations we may enquire whether any useful information can be salvaged by incorporating TRMOM. The failure of Hückel theory stems from the symmetrical distribution of energy levels about the origin $\alpha = 0$ which separates bonding and antibonding levels (Figure 4-8).

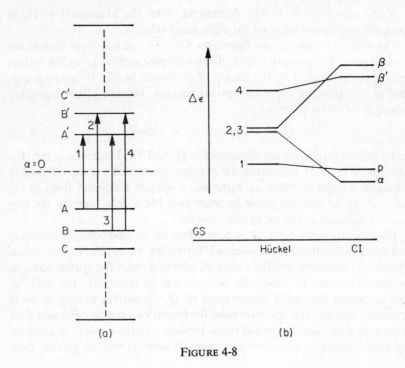

FIGURE 4-8

According to Figure 4-8(a) the transitions of lowest energy involve 'excitations' denoted by arrows 1, 2, 3 and 4, in that order. The transition energies $\Delta\epsilon_2$, $\Delta\epsilon_3$ are equal, due to the symmetry of levels about $\alpha = 0$, thus giving rise to a degenerate condition, shown to the left side of Figure 4-8(b) where the excited 'states' are shown relative to the ground state GS. Hückel theory can go no further, but the CI methods, which are treated in the last chapter, show that the degeneracy is resolved when π-electron-repulsion terms are included explicitly, and the resulting splitting can be large, as represented above CI in Figure 4-8(b).

Characteristic features of the observed ultraviolet (UV) spectra of polycyclic hydrocarbons were first recognized and classified by Clar[13] and subsequently by Platt and Klevens.[14] Three main bands are generally observed, which in order of increasing energy of excitation are the α (weak), p (medium) and β, β' (strong) bonds in Clar's notation, and it is now known that these bonds arise essentially from the interactions amongst excited configurations $\Psi_{exc}(i \to k')$ as represented schematically in Figure 4-8(b). In molecules containing three or more aromatic rings, such as pentacene, for example, the p band may fall below the α band.

Although Hückel theory cannot reproduce the energies of excitation correctly when degeneracies arise, as indicated in Figure 4-8, it is pertinent to enquire whether any useful information relating to spectroscopic problems can be derived from the calculation of transition moments within Hückel theory, by the program TRMOM. However, before discussing results obtained from TRMOM it is necessary to consider the question of units for calculating energies of excitation. In formulating the secular equations in Hückel theory all energies are expressed in units of β, the standard carbon resonance integral, and energy levels themselves are calculated in terms of multiples of β. Some value of β must be assigned, therefore, to match observed spectra, since β is generally interpreted as an adjustable parameter. Because benzene is highly degenerate in the Hückel approximation it is sensible to adopt a value that matches the non-degenerate $\Psi(A \to A')$ transition of naphthalene, which, in accordance with Figure 4-8(a) is to be identified with the p band. This band occurs around 36,000 cm^{-1} and a 'round' figure of $\beta = 3\cdot6$ eV produces (in TRMOM) a corresponding theoretical value. On substituting this value for $\bar{\nu}$ in equation (4-16) and calculating Q for the two-dimensional plane of naphthalene from equation (4-19), assuming a uniform C–C bond length of $1\cdot40$ Å$_z$, the value obtained for the oscillator strength f is $0\cdot52$.

The results quoted in Table (4-2) are obtained directly from the computer program TRMOM, for naphthalene, for the transitions $i \to k'$

4

corresponding to all possible nine 'excitations' involving $i = 3, 4, 5$ and $k' = 6, 7, 8$. These comprise three non-degenerate and three doubly degenerate excitations in the Hückel approximation, and the four transitions $i = 4, 5$ and $k' = 6, 7$ give the excitation energies $\Delta\epsilon_1 = 35{,}900$ cm^{-1}, $\Delta\epsilon_2 = \Delta\epsilon_3 = 47{,}000$ cm^{-1} and $\Delta\epsilon_4 = 58{,}000$ cm^{-1} depicted in Figure 4-8, which, following CI yield the α, p, β and β' bands.

Table 4-2

Naphthalene transitions $\Psi(i \rightarrow k')$

i	k'	$\Delta\epsilon$(cm^{-1})	$m_{ik'}^x$	$m_{ik'}^y$	$f_{\text{theor.}}$
5(A)	6'(A')	35,900	0	0·820	0·5233
5(A)	7'(B')	47,000	1·041	0	1·1048
4(B)	6'(A')	47,000	1·041	0	1·1048
4(B)	7'(B')	58,000	0	0·700	0·6170
5	8'	55,800	0	0	0
3	6'	55,800	0	0	0
4	8'	66,900	0	0	0
3	7'	66,900	0	0	0
3	8'	75,700	0	0·641	0·6752

Energy-level indices: 1–5 occupied, 6'–10' virtual, in ascending order.

Now consider the calculated intensities $f = 1 \cdot 1048$ of the lowest degenerate excitations. Although Hückel theory cannot resolve these energetically, the sum and difference combinations

$$\Psi_1 = \frac{1}{\sqrt{2}}[\Psi(A \rightarrow B') + \Psi(B \rightarrow A')]$$

$$\Psi_2 = \frac{1}{\sqrt{2}}[\Psi(A \rightarrow B') - \Psi(B \rightarrow A')]$$

that correspond respectively to the α and β bands of Figure 4-8(b) yield, on combining the calculated component intensities appropriately

$$f_\alpha = 0$$

$$f_\beta = 2 \cdot 21$$

which match the values obtained by more sophisticated methods of calculation.[15] Indeed the agreement goes further. More advanced methods indicate that the polarizations of absorption for α and β bands are parallel to the x axis, and of the p, β' bands, to the y axis, in accordance with the interpretation of the Hückel intensities given above.

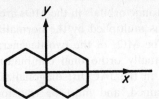

Thus although Hückel theory cannot resolve degeneracies energetically, the method provides valuable information on intensities and polarizations which can be used effectively in identifying spectroscopic states.

The method can be applied in an interesting way for the complex case of benzene where four-fold degeneracies occur for transitions involving the levels A, A′, B and B′. The energy levels and MOs of benzene are well known and can be usefully represented in the following form.

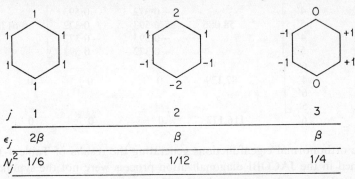

bonding levels and orbitals

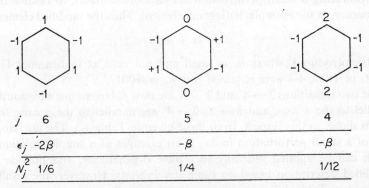

antibonding levels and orbitals

FIGURE 4-9

The coefficients of atomic orbitals in the MOs are the numbers attached to corresponding atoms multiplied by the normalizing factors N_j; ϵ_j are the orbital energies. The MOs of the two degenerate levels $\epsilon = \pm\beta$ are not unique, since mutually orthogonal combinations of each pair are equally valid. In fact, in the JACOBI diagonalization routine arbitrary combinations are obtained, and must be transformed to produce the symmetry orbitals presented above.

The results of Table 4-3 were obtained for benzene from the computer program TRMOM. All nine possible single 'excitations' are included.

Table 4-3

Benzene transitions; Hückel orbitals

i	k'	$\Delta\epsilon(cm^{-1})$	$m_{ik'}^x$	$m_{ik'}^y$	$f_{theor.}$
3	4'		−0·372	0·593	
3	5'	58,086	0·593	0·372	0·6176
2	4'		−0·593	−0·372	
2	5'		−0·372	0·593	
3	6'				
1	4'	87,129	0	0	0
2	6'				
1	5'				
1	6'	116,172	0	0	0

The orthonormal eigenfunctions generating the results of Table 4-3 as obtained in the JACOBI diagonalization process were not the symmetry orbitals (Figure 4-9) of the molecule. These can be produced in practice by introducing a small perturbation (as part of the data), to resolve the degeneracy, in for example, a diagonal element. Thus the modified element

$$\alpha_1 = \alpha + h_1\beta$$

can be introduced, where h_1 is small and not zero as in benzene. The results of Table 4-4 were obtained with $h_1 = 0·001$.

The two transitions $3 \rightarrow 4'$ and $2 \rightarrow 5'$ are now polarized unambiguously parallel to the x axis, and $3 \rightarrow 5'$, $2 \rightarrow 4'$ are parallel to the y axis; the results differ, in this respect, from those given in Table 4-3. The introduction of a small perturbation in $\delta\alpha_1$ is an example of a simple technique which can be applied generally to resolve degeneracies, and provide a simplified description based on symmetry orbitals. However, the results obtained in Table 4-3 are, in principle, more realistic, since the implied arbitrariness in directions of polarization relates correctly to the absence

Table 4-4

Benzene transitions; Hückel symmetry orbitals

i	k'	$\Delta\epsilon(\text{cm}^{-1})$	$m^x_{ik'}$	$m^y_{ik'}$	$f_{\text{theor.}}$
3	4'	58,076	−0·700	0	0·618
3	5'	58,086	0	0·700	0·618
2	4'	58,086	0	0·700	0·618
2	5'	58,095	0·700	0	0·618
3	6'	87,124	0	0	0
1	4'	87,124	0	0	0
2	6'	87,133	0	0	0
1	5'	87,133	0	0	0
1	6'	116,172	0	0	0

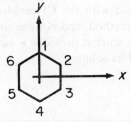

of uniquely specified coordinate axes in the benzene plane; the modification $\delta\alpha_1$ defines axes, and conditions the form of solution obtained.

Now the α, p, β and β' bands are derived from the four lowest, degenerate 'excitations' given in Table 4-4, which shows how the corresponding configurations $\Psi'(i \to k')$ combine in pairs, according to the direction of polarization of the components. These combinations are given in Table 4-5 with assignments, polarizations, and f values. The excitation energies quoted in this table are obtained from an independent CI calculation (Chapter 7) and are included simply to give a complete assignment.

Table 4-5

Benzene transitions: combinations of configurations

Band	Wavefunction	$\Delta\epsilon(\text{cm}^{-1})$	Polarization	f
α	$(3 \to 5') - (2 \to 4')$	39,000	y	0
p	$(2 \to 5') + (3 \to 4')$	43,000	x	0
β'	$(2 \to 5') - (3 \to 4')$	56,000	x	1·23
β	$(3 \to 5') + (2 \to 4')$	56,000	y	1·23

Thus, although Hückel theory cannot calculate excitation energies $\Delta\epsilon$ correctly, it obviously is capable of providing useful information on intensities and polarizations which can be related to the theoretical description of spectroscopic states.

It is interesting to pursue the problem further by introducing modifications to coulomb integrals which can apply to pyridine and other N derivatives of benzene. It is possible to trace, at least approximately, changes in component intensities from which changes in the α, p and β, β' bands can be deduced. It will be noted, for example, that both the α and p bands become allowed, but with small intensities, and that since the UV spectra of pyridine and other N derivatives of benzene do not differ dramatically from that of the parent molecule, the CI comparable to that in benzene must operate.

It is obvious, from the previous discussion, that the calculations obtained from TRMOM within Hückel theory, begin to encroach upon territory usually associated with the CI problem (Chapter 7). This is a valuable property of the method, and not one to be discarded because the description is incomplete, since it provides a useful account of the nature of the CI problem and of its solution.

4.5 $d_\pi-p_\pi$ BONDING

The properties of molecules containing atoms with valence shell d-orbitals provide, in certain circumstances, evidence of conjugation effects involving $d_\pi-p_\pi$ bonding. The most celebrated examples are the phosphonitrilic halides which were studied originally by Craig and Paddock,[16,17,18] and later by Dewar and his associates[19]; both investigations were carried out in terms of the Hückel method, and conflicting conclusions were reached about the nature of the bonding. In both cases group theoretical methods were used in setting up the Hückel secular equations describing π-bonding, to simplify the algebraic processes and to clarify the orbital description. The same problems can, however, be solved by the standard computer programs, by following the customary practice of constructing an incidence matrix NUCK which specifies appropriate interactions between neighbouring atoms.

Although these molecules fall outside the class of systems generally considered for inclusion here, they provide a particularly interesting example of the value of the computational approach both for problem solving, and as an aid to resolving questions concerning the interpretation of theoretical models. In particular, solutions obtained by computer methods demonstrate that a unique MO description of the bonding cannot be defined, and that Dewar's model, which is described in terms of localization within P–N–P 'islands' is, nonetheless, equivalent to conjugation embracing the complete hexagonal ring.

The simplest ring molecule of the phosphonitrilic halide series is the trimer with a six-membered ring composed of PN bonds of equal length,

$$
\begin{array}{c}
\text{Cl} \quad \text{Cl} \\
\text{P} \\
\text{N} \quad \text{N} \\
\text{Cl} - \text{P} \qquad \text{P} - \text{Cl} \\
\text{N} \\
\text{Cl} \qquad \text{Cl}
\end{array}
$$

which can, for theoretical purposes, be assumed planar, in the first instance, and comparable in structure, therefore, to benzene. The two chlorine atoms attached to each phosphorus atom lie symmetrically above and below the plane of the ring.

Let local coordinates be constructed at each P atom, with the z axis perpendicular to the plane, and the y axis bisecting the $N\hat{P}N$ angle. Then the d-type atomic orbitals that are, in principle, available for π-bonding in the six-membered $(PN)_3$ ring are the $3d_{xz}$ and $3d_{yz}$ orbitals. These have similar 'shapes' with axes lying in perpendicular planes, but are characteristically different in the way they overlap adjacent $N(2p_z)$ atomic orbitals. Whereas d_{yz} orbitals match $N(p_z)$ orbitals in the regions of effective overlap, d_{xz} match on one side of the local yz plane, and mismatch on the other (Figure 4-10). Resonance integrals between d_{yz} and p_z are, therefore,

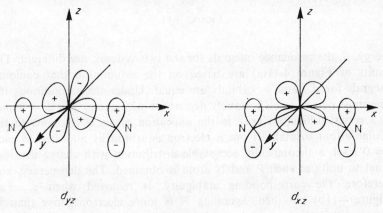

FIGURE 4-10

always negative, and between d_{xz} and p_z positive on one side and negative on the other side of the local yz plane.

In the model proposed by Craig and Paddock[19-21] for describing $d_\pi-p_\pi$ bonding in $(PN)_3Cl_6$ and similar molecules, $P(3d_{yz})$ orbital participation in conjugation was ignored, and π-electron MOs were constructed as linear combinations of three $P(3d_{xz})$ and three $N(2p_z)$ atomic orbitals. The reasons for neglecting d_{yz} orbital participation were based upon the influence of the adjacent Cl ligands on d-orbital size and effective electronegativity, and are explained in the original paper[20]; they do not concern the subsequent calculation of $d_{xz}-p_z$ π-bonding. The energy-level diagrams in Figure 4-11(a, b) for the six-membered $(PN)_3$ ring determined by this model, show an 'inversion' with respect to the form of energy levels for benzene in Figure 4-11(c), and these results can easily be confirmed by computer calculations, as described later. Obviously the energy scales for $(PN)_3$ and for benzene in Figure 4-11 are different since the units of

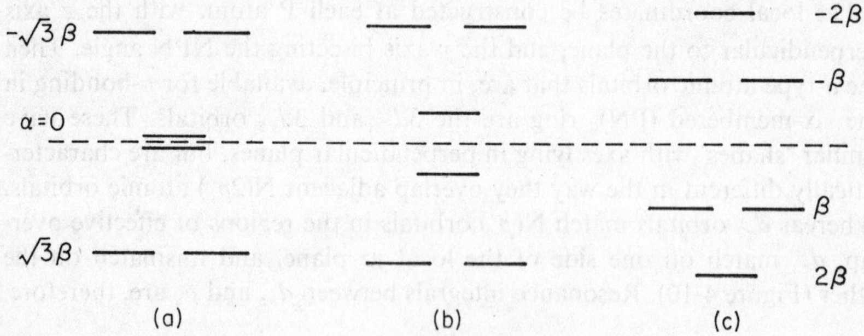

FIGURE 4-11

energy, β, the resonance integrals for the two systems, are different. The results of Figure 4-11(a) are based on the assumption that coulomb integrals for d_{xz} and p_z orbitals are equal. Under these conditions the innermost pair of levels is doubly degenerate with energy $\epsilon = \alpha = 0$ and, as a result, difficulties arise in the allocation of six π electrons to the available four orbitals. If one π electron is arbitrarily allocated to each $\epsilon = 0$ level, a theoretically acceptable distribution with charge densities equal to unity at each P and N atom is obtained. The degeneracy, and therefore, the corresponding ambiguity, is removed when $\alpha_{xz} \neq \alpha_z$ (Figure 4-11b) but then, assuming N is more electronegative than P, charge distributions are obtained in which $q_N \sim 1\frac{1}{3}$ and $q_P \sim \frac{2}{3}$. This uneven charge distribution persists when the difference in coulomb integrals is small, but not zero, and the description is then theoretically unrealistic. This problem provides a further manifestation of the failure of Hückel theory in describing degenerate situations where the allocation of electrons to orbitals is ambiguous, and any interpretation based on this kind of solution must be suspect. Ambiguity in the occupation of the lowest degenerate pair does not arise since four π electrons are available to complete the shell. Craig and Paddock confirmed the existence of delocalization, or π-type bonding in d_{xz}–p_z systems, analogous to that in p_z–p_z systems but with the inversion of energy levels as indicated in Figure 4-11.

The model proposed by Dewar and his associates differs from that of Craig and Paddock by the inclusion of $P(d_{yz})$ orbitals which were combined in sum and difference forms with $P(d_{xz})$ orbitals to give a pair of d_π type orthonormal orbitals of the form

$$d_\pi^a = \frac{1}{\sqrt{2}}(d_{yz} + d_{xz}); \qquad d_\pi^b = \frac{1}{\sqrt{2}}(d_{yz} - d_{xz}) \qquad (4\text{-}21)$$

for each P atom. Each d_π orbital overlaps efficiently p_z orbitals of just one of the two adjacent N atoms, as indicated in Figure 4-12 by the disposition of the axes a and b relative to the P–N bond directions. The

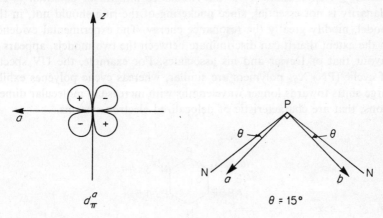

FIGURE 4-12

d_π–p_π problem is formulated in Dewar's model in terms of a resonance integral

$$\beta^* = \int\phi(d_\pi^\nu)h_\pi\phi(p_z)\,d\tau \ (\nu = \text{a, b}) \qquad (4\text{-}22)$$

which by reason of the combination (4-21) is $\sqrt{2}$ times the resonance integral β used in Craig and Paddocks results (Figure 4-11). For simplicity an idealized model is chosen, in the first instance, in which all coulomb integrals are taken to be equal. The bonding is then described in terms of three independent and equivalent three-centre P–N–P bonds, allylic in character, each accommodating two π electrons. Conjugation is interrupted at each P atom, since the transmission of delocalization effects depends primarily upon overlap of d_π^a, d_π^b orbitals, which is zero.

A general $(PN)_n$ ring system can, therefore, be visualized in terms of a σ-bonded framework with localized P–N–P π-type bonding superposed, and aromatic character, which depends upon delocalization throughout the framework, is accordingly vitiated. The energy levels of the system $(PN)_3$ are equivalent to those for allyl, repeated three times, namely $\epsilon = 0$, $\pm\sqrt{2}\beta^*$, where β^* is the resonance integral of equation (4-22). Planarity is not essential, since puckering of the ring should not, in this model, modify greatly the resonance energy. The experimental evidence, to the extent that it can discriminate between the two models, appears to favour that of Dewar and his associates. For example, the UV spectra of cyclic $(PN)_nX_{2n}$ polymers are similar, whereas cyclic polyenes exhibit large shifts towards longer wavelengths with increase in molecular dimensions, that are characteristic of delocalized electron systems.

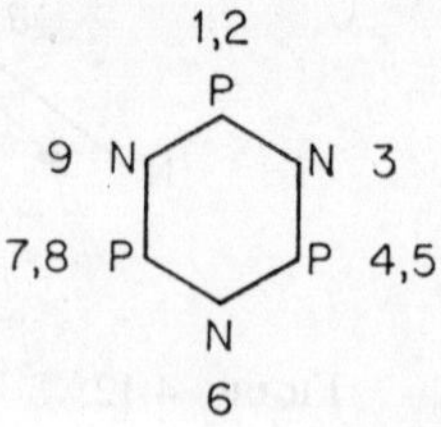

$$
\begin{array}{c}
1,2 \\
P \\
9 \ N \quad\quad N \ 3 \\
7,8 \ P \quad\quad P \ 4,5 \\
N \\
6
\end{array}
$$

FIGURE 4-13

We can now return to explore the nature of the π-bonding as described by the solutions emerging from computer calculations. The atomic orbitals are labelled according to the sequence shown in Figure 4-13 where labels 1, 4, 7 identify $P(d_{yz})$ orbitals, 2, 5, 8 identify $P(d_{xz})$ and 3, 6, 9 identify $N(p_z)$ orbitals. An idealized model, in which all coulomb integrals are taken to be equal, is adopted in the first instance, as in Craig and Paddock's and Dewar's models. The incidence matrix NUCK defining the molecule is then written in the usual way.

```
1  0
2  0 0
3  1 1 0
4  0 0 1 0
5  0 0 1 0 0
6  0 0 0 1 1 0
7  0 0 0 0 0 1 0
8  0 0 0 0 0 1 0 0
9  1 1 0 0 0 0 1 1 0
```

Each off-diagonal element 1 is formally changed to $-1\cdot0$ in the subroutine INPT so that energy levels are computed in units of β, the resonance integral, assumed to be the same for both d-type orbitals. This representation must be changed to $+1\cdot0$ in the modification subroutine MODH for resonance integrals referring to mismatched d_{xz}, p_z orbitals, the appropriate modifications being

$$0 \quad 3 \quad 0 \quad 2 + 01\cdot000$$
$$0 \quad 6 \quad 0 \quad 5 + 01\cdot000$$
$$0 \quad 9 \quad 0 \quad 8 + 01\cdot000$$

The computer solution obtained from this specification gives nine energy levels, composed of three triply degenerate levels, $\epsilon_1, \epsilon_2, \epsilon_3 = 2\beta$; $\epsilon_4, \epsilon_5, \epsilon_6 = 0$; and $\epsilon_7, \epsilon_8, \epsilon_9 = -2\beta$. The six available π electrons occupy the triply degenerate bonding levels, $\epsilon_{1,2,3} = 2\beta$ in pairs with coupled spins. The charge density at each atom is unity, but is 'shared' equally at the P atoms between d_{xz} and d_{yz} orbitals,

$$q_N = 1; \qquad q_P(d_{xz}) = 0\cdot5; \qquad q_P(d_{yz}) = 0\cdot5$$

The bond orders are modulus unity in each bond, but 'shared' between $d_{xz} - p_z$ and $d_{yz} - p_z$ so that

$$p(d_{xz} - p_z) = \pm0\cdot5; \qquad p(d_{yz} - p_z) = 0\cdot5$$

When adjacent d_{xz}, p_z orbitals mismatch, corresponding bond orders are $-0\cdot5$, and the contribution to the total π-electron energy is obtained, for these bond orders, by multiplying by $-\beta$. The ground-state distribution may be represented in the usual way by charge densities and bond orders attached to atoms and bonds (Figure 4-14).

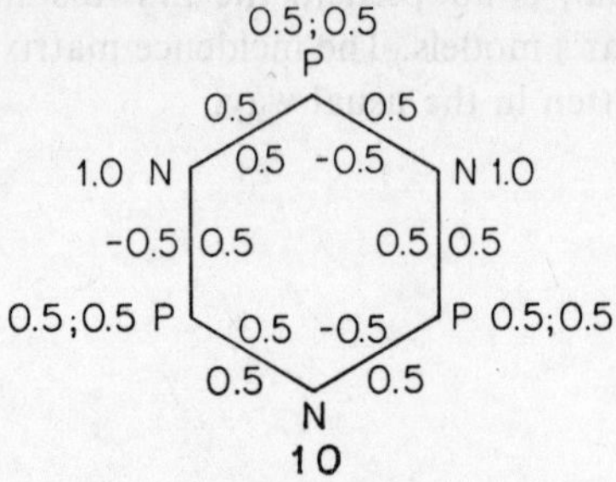

FIGURE 4-14

Although the system of energy levels obtained from the computer solution is the same as that calculated by Dewar, since, by equation (4-21, 22), $\beta^* = \sqrt{2}\beta$ the MOs themselves are different. Indeed, the form of MOs for the degenerate situation, as they emerge from the computer solution, cannot be anticipated, since it may depend upon the numbering system used in formulating the NUCK matrix, and on various factors, including the word length, involved in the processing during diagonalization. This arbitrariness in the form of MOs for degenerate situations creates problems of identification, but has the merit of demonstrating that the orbital description is not unique. In any case, it is always possible to transform the MOs to a form which provides a 'better' description of the molecule; in particular, the 'localized' MOs of the P–N–P islands of Dewar's model can be obtained in this way, though, again, this description is not unique.

Another important description is that which reflects the full symmetry of the $(PN)_3$ hexagon. It is possible to remove the degeneracies described above and to extract the symmetry orbitals from computer solutions by applying small modifications to diagonal elements similar to those introduced to remove degeneracies in benzene, as described in Section 4-4. If we denote by Φ_j ($j = 1, 2, \ldots 9$) the MOs corresponding to the levels ϵ_j which reflect the hexagonal symmetry of the molecule, then one pair, Φ_1 and Φ_9 involves $N(p_z)$ and $P(d_{yz})$ atomic orbitals only, and is obtained in the form

$$\Phi_1^+, \Phi_9^- = \frac{1}{\sqrt{6}}(\phi_1 \pm \phi_3 + \phi_4 \pm \phi_6 + \phi_7 \pm \phi_9)$$

where the superfix simply identifies the sign to be taken in the MO. These two orbitals have precisely the same analytical form as the pair of MOs of benzene (Section 4-4) which describe the lowest bound and the highest unoccupied levels with energies $\epsilon = \pm 2\beta$ (Figure 4-9). The remaining seven MOs are rather more difficult to interpret, but can be conveniently expressed in terms of group 'orbitals' for P_3 and N_3, as indicated below. Although the introduction of these group orbitals begins to create an impression that group theoretical methods are implicit in deriving the solution, this is not the case. The group orbitals simply provide a compact notation for describing and identifying the MOs, which are obtained in numerical form from the computer solution, with no emphasis on grouping.

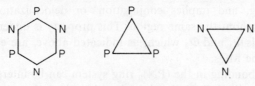

FIGURE 4-15

Orthonormal group functions for an equilateral triangle carrying equivalent atomic orbitals at each apex, may be written in the form

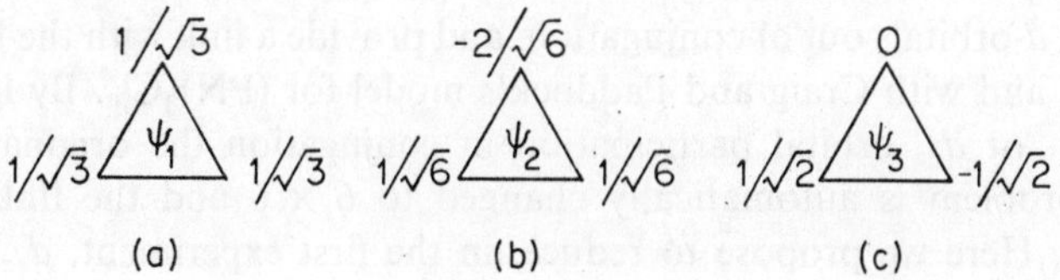

FIGURE 4-16

These apply to each of the three atomic-orbital systems $P_3(d_{xz})$, $P_3(d_{yz})$ and $N_3(p_z)$ independently. The identification $(a, b, c) \equiv (\psi_1, \psi_2, \psi_3)$ will be made, the $N(p_z)$ set of ψs being unprimed, $P(d_{yz})$ primed once, and $P(d_{xz})$ primed twice. Then the remaining molecular orbitals of the $(PN)_3$ π-electron system can be written in the form

$$\Phi_2^+, \Phi_7^- = \frac{1}{\sqrt{2}}\psi_2 \pm \frac{1}{2\sqrt{2}}\psi_2' \pm \sqrt{\frac{3}{8}}\psi_3''$$

$$\Phi_3^+, \Phi_8^- = \frac{1}{\sqrt{2}}\psi_3 \pm \sqrt{\frac{3}{8}}\psi_2'' \pm \frac{1}{2\sqrt{2}}\psi_3'$$

with three orbitals describing the triply degenerate level lying in the zero of energy $\epsilon_4 = \epsilon_5 = \epsilon_6 = 0$

$$\Phi_4 = \psi''_1$$

$$\Phi_5 = \tfrac{1}{2}\psi''_2 + \frac{\sqrt{3}}{2}\psi'_3$$

$$\Phi_6 = \tfrac{1}{2}\psi''_3 - \frac{\sqrt{3}}{2}\psi'_2$$

The form of the MOs Φ_j ($j = 1, 2, \ldots 9$) is, for present purposes, of importance because it emphasizes that each orbital embraces the six-membered ring, and implies conjugation, or delocalization of the π electrons throughout the same region. This property is underlined by the familiar orbitals Φ_1 and Φ_9 which, as indicated above, are equivalent in form to benzene MOs.

Thus $d_\pi-p_\pi$ bonding in the $(PN)_3$ ring system can be interpreted either in terms of localized P–N–P 'islands' with, effectively, no conjugation effects operating between islands, as suggested by the model due to Dewar, or, alternatively, as fully conjugated systems. The two interpretations are equivalent and complementary.

We propose now to outline a set of experiments which have the effect of taking d-orbitals out of conjugation, and provide a link with the benzene solution, and with Craig and Paddock's model for $(PN)_3Cl_6$. By ignoring either d_{yz} or d_{xz} orbital participation in conjugation the original 9×9 matrix problem is automatically changed to 6×6 and the link is not achieved. Here we propose to reduce, in the first experiment, d_{xz} orbital participation progressively, within the 9×9 matrix problem, by increasing systematically the three coulomb integrals

$$\alpha_{xz} = \alpha + h_{xz}\beta$$

where α is the coulomb integral common to the remaining $P(d_{yz})$ and $N(p_z)$ atomic orbitals, β the common resonance integral, and h_{xz} is made progressively negative. The discussion given earlier in Section 4-2 indicates that when the three coulomb integrals α_{xz} are made increasingly electropositive, three levels emerge eventually above the 'band', and the corresponding MOs become increasingly 'localized' as $P_3(d_{xz})$ group orbitals. The remaining $P(d_{yz})$ and $N(p_z)$ orbitals participate in conjugation which increasingly resembles, in character, the conjugation found for benzene. Thus the six energy levels, excluding the three levels that emerge above the band change towards the limiting values $\pm 2\beta$, $\pm\beta$ and $\pm\beta$, which characterize the benzene system (Figure 4-17). The levels $\epsilon_1 = 2\beta$ and $\epsilon_9 = -2\beta$ and the corresponding 'benzene-like' MOs Φ_1 and Φ_9 of the original d_π–p_π

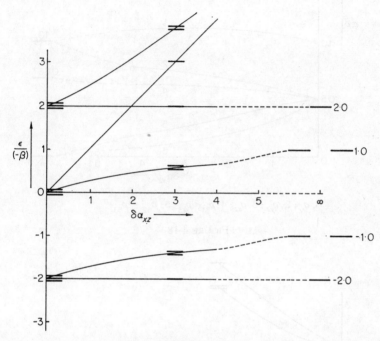

FIGURE 4-17

system, in fact remain unchanged throughout this process. At the same time charge densities and bond orders also manifest changes towards a benzene-type configuration with charge densities $q_N = q_{yz} = 1$, and $p_{NP} = 2/3$ (Figure 4-18).

In a second experiment d_{yz} orbital participation in bonding is diminished, in a similar manner, by increasing progressively the three-coulomb integrals

$$\alpha_{yz} = \alpha + h_{yz}\beta$$

In this case three levels associated increasingly with $P_3(d_{yz})$ orbitals only, emerge above the 'band', and the remaining levels change towards those obtained by Craig and Paddock, namely $\pm\sqrt{3}\beta$, $\pm\sqrt{3}\beta$, 0, 0 as limiting values (Figure 4-19).

band three-center delocalized ... suppose 2C ... be valid ... with electrons the benzene system (1) ... 1,4 ... or 1,3 (3) ... the levels ... are degenerate ... and the corresponding resonance ... are 0 and ... ($\sqrt{3}$) ... of each ...

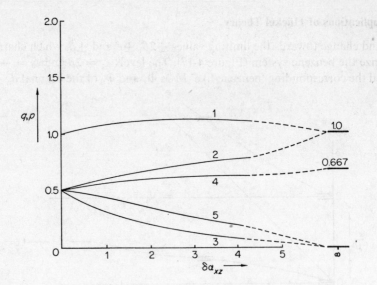

1, q_N; 2, q_{yz}; 3, q_{xz}; 4, $\rho_{31} = \rho_{43}$; 5, $\rho_{32} = -\rho_{53}$

FIGURE 4-18

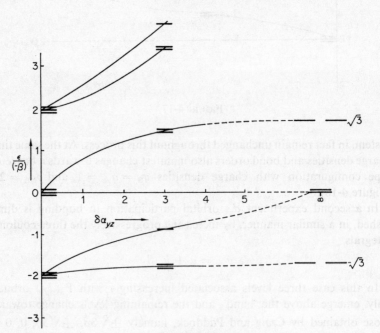

FIGURE 4-19

system, in fact remain unchanged throughout this process. As the ... the charge densities and bond orders also ... of changes ... to ... type configuration, with charge densities $q_N = \ldots = \ldots$ ... (Figure 4 ...

In a second experiment ... orbital ... bond order is diminished, in a similar manner, ... the bond order ... integrals

In this case, three levels, associated ... energy ... with $\Gamma \ldots$ obtain ... only, emerge above the level, and the remaining levels shift ... those obtained by Coffey and Feldman, namely ... so ... 0 as limiting values (Figure 4-19).

The changes in charge densities and bond orders may be represented in the form of Figure 4-20.

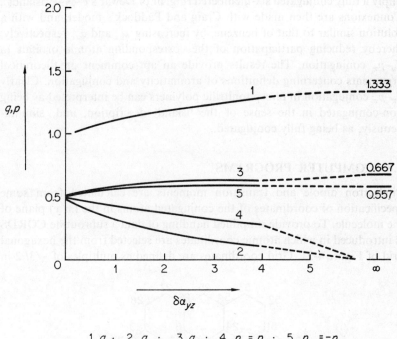

1, q_N;   2, q_{yz};   3, q_{xz};   4, $\rho_{31} = \rho_{43}$;   5, $\rho_{32} = -\rho_{53}$

FIGURE 4-20

The limiting values $q_N = 1\frac{1}{3}$, $q_{xz} = \frac{2}{3}$ do not represent a satisfactory description of the charge distribution, and are due to the failure of Hückel theory to deal adequately with the assignment of π electrons in the neighbourhood of the degeneracy $\epsilon = 0$ which also applies in Craig and Paddock's model (Figure 4-11) as described earlier. If one π electron is arbitrarily assigned to each of the two levels which coincide in the degenerate situation, an acceptable even charge distribution with $q_N = q_{xz} = 1$ is obtained, but a small perturbation recreates the anomaly. The origin of the problem and details of the discontinuity in the charge distributions can best be studied by examining the corresponding orbital and energy-level changes in these experiments.

The calculations described above, in which parameters of the equations are varied, are made possible in a practical sense largely by the availability of high-speed computational methods. Both sets stem from a model in which all coulomb integrals are equal, $\alpha_{xz} = \alpha_{yz} = \alpha_z$, and which can be described in terms of the symmetry orbitals Φ_j ($j = 1, 2, \ldots 9$) which imply a fully conjugated six-membered ring, or of Dewar's P–N–P 'islands'. Connexions are then made with Craig and Paddock's model, and with a solution similar to that of benzene, by increasing α_{yz} and α_{xz} respectively, thereby reducing participation of the corresponding atomic orbitals in d_π–p_π conjugation. The results provide an apt comment on theoretical arguments concerning definitions of aromaticity and conjugation. Clearly d_π–p_π conjugation in phosphonitrilic polymers can be interpreted as being non-conjugated in the sense of the 'island' description, and, simultaneously, as being fully conjugated.

4.6 COMPUTER PROGRAMS

Pi-electron dipole and transition moments are calculated from some specification of coordinates of the conjugated atoms in the (x, y) plane of the molecule. To provide simplified handling of data a subroutine CORDS is introduced in which atomic coordinates are selected from the hexagonal grid of Figure 4-21. Grid coordinates are defined as multiples of $\sqrt{3}l/2$ in

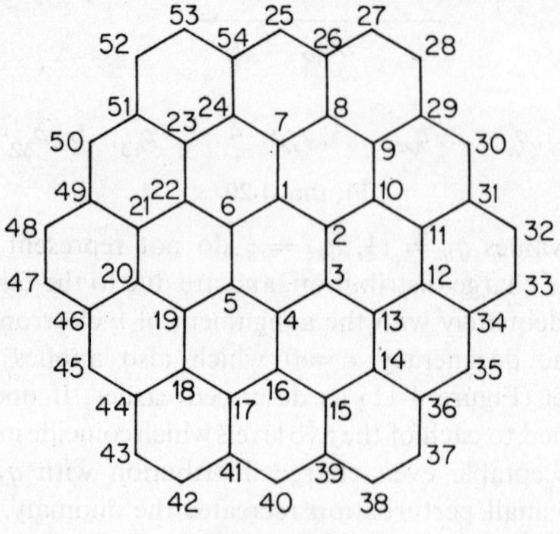

FIGURE 4-21

the x direction and as multiples of $l/2$ in the y direction, where l is the (uniform) bond length. Atom 1 has grid coordinates $(0, 2)$, atom 2 is $(1, 1)$, atom 3 $(1, -1)$, atom 4 $(0, -2)$ and so on, and the complete set of grid coordinates, given in Table 4-6 for the 54 atoms represented in Figure 4-21 is read, in FORMAT (40I2), into the arrays IX, IY so that corresponding elements identify a grid point.

Table 4.6

Data cards for the arrays IX, IY

First card for IX (40 atoms)

+0+1+1+0−1−1+0+1+2+2+3+3+2+2+1+0−1−2−2−3−3−2
−2−1+0+1+2+3+3+4+4+5+5+4+4+3+3+2+1+0

Second card for IX (14 atoms)

−1−2−3−3−4−4−5−5−4−4−3−3−2−1

First card for IY

+2+1−1−2−1+1+4+5+4+2+1−1−2−4−5−4−5−4−2−1+1+2
+4+5+8+7+8+7+5+4+2+1−1−2−4−5−7−8−7−8

Second card for IY (14 atoms)

−7−8−7−5−4−2−1+1+2+4+5+7+8+7

Atom coordinates (x_i, y_i) are computed within the subroutine CORDS by multiplying grid coordinates for selected molecular frameworks by $\sqrt{3}EL/2$ and $EL/2$, where $EL = 1.40$ is taken to be the uniform bond length in Angstrom units. Since the hexagonal grid may be required for several successive calculations, involving different molecular frameworks, input of grid coordinates IX, IY conveniently precedes initiation of the outer loop in the main program.

The use of the hexagonal grid of Figure 4-21 is fairly obvious. Naphthalene, for example, can be defined by the following sequence of numerical labels

10 11 12 13 03 04 05 06 01 02

or by

25 26 08 07 24 23 51 52 53 54

and so on. The chosen set of N labels, where N is the number of conjugated atoms, is read under FORMAT (40I2) into the array NATM and the atom coordinates are processed by extracting corresponding IX and IY values. The sequence in the chosen set is immaterial at this point. It is important, however, that the incidence matrix NUCK that defines the bonding in subroutine INPT should match the sequence. For example, both sequences for naphthalene match the following bonding scheme for NUCK but the sequence

54 24 23 51 52 53 25 26 08 07

represents a reordering of the second set, that matches the NUCK bonding scheme

The same NUCK would match the 'skew' sequence for naphthalene

24 07 01 06 22 23 54 25 26 08

though computed values of x and y components of dipole and transition moments obtained from the programs described below would be different. Resultant moments are always the same in magnitude, and in directions of polarization referred to molecular axes, when different basic frameworks are chosen, though components and polarizations referred to grid co-ordinates may be different.

At first sight, specification of a molecule in terms of hexagonal grid coordinates should dispense with the framework definition given by the incidence matrix NUCK. However, ambiguous situations can arise from a grid definition, for example, as between *cis*-stilbene and phenanthrene

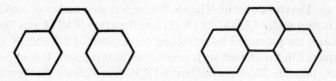

and these are resolved by the NUCK specification.

DIMO (p. 117)

The argument list of the subroutine DIMO that calculates the dipole moment due to the π-electron system, contains N, the number of conjugated atoms, the arrays X and Y that hold the coordinates (x_i, y_i) of the $i = 1, 2, \ldots$ N atoms, and PRS, the bond-order matrix. The subroutine calculates and prints EMU, the π-electron dipole moment, and components SX, SY parallel to the coordinate axes. The dipole moment is calculated in accordance with the overlap approximation which takes into account diagonal elements PRS(I, I) only of the charge distribution.

Additional statements are introduced in MODH to allow modification of elements of the array Z(I) of effective nuclear charges from the initial values Z(I) = 1 set in PAHY. These instructions constitute the loop initiated by statement 22 in MODH. Dipole moments are then calculated from the net charges at each conjugated atom

$$P = PRS(I, I) - Z(I)$$

TRMOM (p. 118)

The subroutine TRMOM calculates transition moments and oscillator strengths for single 'excitations' between specified π-electron energy levels. It contains a program segment for the automatic selection of levels that is primarily intended to simplify input data, and can be applied to parent AHs and related modified molecules. The significant transitions are those of lowest energies involving 'excitations' amongst levels that lie immediately on either side of the zero α of energy. The subroutine argument LVLS contains, on entry, a preassigned integral value that specifies the levels to be taken into account; for example, if LVLS = 3 all transitions between the three highest occupied and the three lowest unoccupied levels will be computed. For naphthalene this specification computes the transitions $(i \rightarrow k')$ in which $i = 3, 4, 5$ and $k' = 6, 7, 8$ and for anthracene $i = 5, 6, 7$ and $k' = 8, 9, 10$. The total number of transitions selected by LVLS is stored in MINK, and is notionally given by MINK = $(LVLS)^2$.

The argument list of the subroutine contains N, the number of conjugated atoms, LVLS the specification for selecting levels, C, the two-dimensional array of MOs, ADIAG, the array of energy levels, and X, Y, the arrays of atomic coordinates. The subroutine computes and prints for each transition

  LL — lower index i
  LH — higher index k'
  TERM — excitation energy

XMIK — $m_{ik'}^{x}$, x component of the transition moment.

YMIK — $m_{ik'}^{y}$, y component of the transition moment.

OSC — oscillator strength

TERM and OSC are calculated with the formal introduction of $\beta = 3.6$ eV in the FORTRAN statement SBETA = 3·6, which is chosen to match the observed transition energy of the p-band of naphthalene, as described earlier.

MAIN program (p. 116)

A modified MAIN program that reads the grid coordinates IX, IY and later calls DIMO and TRMOM is presented.

Additional notes

The hexagonal grid cannot be used for non-alternants like azulene and certain other molecules, though most alternants with mesomeric groups attached can conveniently be referred to a hypothetical parent alternant. Nitrobenzene can, for example, be associated with the sequence

10   11   12   13   03   02   09   08   29

and a corresponding incidence matrix NUCK which generates a HUCK matrix that can be modified by the introduction of appropriate parameters for the NO_2 group in MODH.

It will be necessary, when the hexagonal grid does not apply, to replace CORDS by a routine reading (x, y) coordinates directly, and to omit input of the grid arrays IX, IY.

Similarly, the simplified specification of LVLS must be used with care in certain situations, though it will always operate, provided the levels called for are available. If replacement is necessary, a program segment reading the indices i of occupied orbitals into the array LL, and corresponding indices k' of unoccupied orbitals in the array LH, must be provided to replace the selection routine in TRMOM that precedes initiation of the loop that begins DO 18.

The program segments concerned with the hexagonal grid and with the LVLS routine are introduced simply to reduce the quantity of input data. They are easy to use, effective in practice, and reduce the risk of error.

Data Set 1

The following set of data will compute π-electron energy levels and orbitals, charge densities, bond orders, free valences, atom–atom polarizabilities, dipole moments, excitation energies, components of transition moments and oscillator strengths for

      (a) naphthalene, quinolene and isoquinolene

and   (b) benzene, pyridine and pyrazine

using the parameter $\delta\alpha_N = 0.5\beta$.

First card for IX ⎤	
Second card for IX grid coordinates (40I2)	
First card for IY	
Second card for IY ⎦	

0002   NMOLS (number of parent hydrocarbons)

0005   LVLS
010005   N, M

```
0
1 0
0 1 0
0 0 1 0                     NUCK
0 0 0 1 0                   (80I1)
0 0 0 0 1 0
0 1 0 0 0 1 0
0 0 0 0 0 0 1 0
0 0 0 0 0 0 0 1 0
1 0 0 0 0 0 0 0 1 0
```

```
10 11 31 32 33 34 12 13 03 02 NATM (40I2)
0003                 NDER
0000                 NMOD     hydrocarbon
0001                 NMOD     quinolene
0101 − 0·500         I, J, X
0000                 IJ, X    Z modification
0001                 NMOD
0202 − 0·500         I, J, X  isoquinolene
0000                 IJ, X    Z modification
0003                 LVLS
006003               N, M
```

```
0
1 0
0 1 0                    NUCK
0 0 1 0                  (80I1)
0 0 0 1 0
1 0 0 0 1 0
```

```
06 22 23 24 07 01              NATM (40I2)

0003                     NDER
0000                     NMOD    hydrocarbon
0001                     NMOD
0101 − 0·500             I, J, X   pyridine
0000                     IJ, X
0002                     NMOD    pyrazine
0101 − 0·500             I, J, X
0404 − 0·500             I, J, X
0000                     IJ, X
```

Data Set 2

The following data set computes the quantities listed in Data Set 1 for aniline. Two calculations are performed with the parameter sets

(a) $\delta\alpha_N = 1·0\beta$, $\beta_{C-N} = 0·7\beta$, $Z_7 = 2$

(b) $\delta\alpha_N = 1·5\beta$, $\beta_{C-N} = 0·7\beta$, $Z_7 = 2$

```
IX, IY cards                     grid coordinates
0001                     NMOLS

0003                     LVLS
007004                   N, M
```

```
0
1 0
0 1 0                    NUCK
0 0 1 0                  (80I1)
0 0 0 1 0
1 0 0 0 1 0
1 0 0 0 0 0 0
```

07 24 54 25 26 08 01   NATM (40I2)

0002	NDER
0002	NMOD
0701 — 0·700	β_{C-N}
0707 — 1·000	$\delta\alpha_N$
072·0	IJ, X $Z_7 = 2$
0000	
0002	NMOD
0701 — 0·700	β_{C-N}
0707 — 1·500	$\delta\alpha_N$
072·0	IJ, X $Z_7 = 2$
0000	

It should be noted that the technique for selecting 'excitations' automatically, according to the value of LVLS, omits, in the present case,
three excitations involving transfers from the lowest bound level ϵ_1 to the
three antibonding levels $\epsilon_{k'}$ ($k' = 5, 6, 7$). The calculation is, in fact,
confined to the nine transitions $\psi(i \to k')$ in which $i = 2, 3, 4$ and $k' =$
5, 6, 7.

The results obtained from Data Set 1 can be compared with those given
in the text in Section 4. A section of this output describing the solution
for pyridine precedes the program listings.

A. Results

```
MOLECULE NO.  1
0
1 0
0 1 0
0 0 1 0
0 0 0 1 0
1 0 0 0 1 0


MODIFICATIONS
 1  1   -0.500
ENERGY LEVELS
   J=    1        2        3        4        5        6

     -2:1074  -1.1672  -1.0000   0.8410   1.0000   1.9337

HÜCKEL ORBITALS
   J=    1        2        3        4        5        6
 1  0.520706 -0.571374  0.000000  0.545913  0.000000 -0.323073
 2  0.418504 -0.190609 -0.500000 -0.366024 -0.500000  0.393128
 3  0.361268  0.348897 -0.500000 -0.238101  0.500000 -0.437110
 4  0.342849  0.597839  0.000000  0.566258 -0.000000  0.452102
 5  0.361268  0.348897  0.500000 -0.238101 -0.500000 -0.437110
 6  0.418504 -0.190609  0.500000 -0.366024  0.500000  0.393128

TOTAL PI-ELECTRON ENERGY =   -8.5493

CHARGE DENSITIES

    1:1952   0.9230   1.0045   0.9499   1.0045   0.9230

FREE VALENCES

    0:4247   0.4090   0.3977   0.4022   0.3977   0.4090

BOND-ORDER MATRIX

    1:1952   0.6537   0.9230  -0.0225   0.6694   1.0045  -0.3261   0.0591

    0:6649   0.9499  -0.0225  -0.3306   0.0045   0.6649   1.0045   0.6537

   -0:0770  -0.3306   0.0591   0.6694   0.9230

DIPOLE MOMENT= 1.0936   XMU=  0.0000   YMU= 1.0936
```

```
TRANSITION MOMENTS

  I    K    ENERGY      XINT      YINT      OSCS
  3    4   53467.05    0.732     0.000     0.6225
  3    5   58086.00    0.000     0.700     0.6176
  2    4   58322.86    0.000    -0.697     0.6142
  2    5   62941.81    0.654     0.000     0.5844
  3    6   85202.82    0.053    -0.000     0.0053
  1    4   85630.62    0.000     0.032     0.0019
  2    6   90058.63   -0.000    -0.011     0.0003
  1    5   90249.57   -0.069     0.000     0.0094
  1    6  117366.38   -0.000    -0.001     0.0000


ATOM-ATOM POLARIZABILITIES

ATOM    1
-0.3754  0.1477 -0.0084  0.0969 -0.0084  0.1477

ATOM    2
 0.1477 -0.3991  0.1588 -0.0058  0.1004 -0.0020

ATOM    3
-0.0084  0.1588 -0.3976  0.1561 -0.0094  0.1004

ATOM    4
 0.0969 -0.0058  0.1561 -0.3974  0.1561 -0.0058

ATOM    5
-0.0084  0.1004 -0.0094  0.1561 -0.3976  0.1588

ATOM    6
 0.1477 -0.0020  0.1004 -0.0058  0.1588 -0.3991

   END


REND;
TIME =  0001  29
    A
```

B. Listings

```
C      HÜCKEL CALCULATIONS
       DIMENSION X(96),Y(96),IX(96),IY(96)
       DIMENSION A(30,30),ADIAG(30),U(30,30),PRS(30,30),FV(30)
       DIMENSION Z(30)
       READ(7,98)(IX(I),I=1,54)
       READ(7,98)(IY(I),I=1,54)
    98 FORMAT(40I2)
       READ(7,99)NMOLS
    99 FORMAT(I4)
       DO 10 KMOLS=1,NMOLS
       WRITE(2,100)KMOLS
   100 FORMAT(1H1,13H MOLECULE NO.,I4)
       READ(7,99)LVLS
       CALL INPT(N,M,A)
       CALL CORDS(N,X,Y,IX,IY)
       READ(7,99)NDER
       DO 10 KDER=1,NDER
       CALL PAHY(N,A,ADIAG,Z)
       CALL MODH(N,A,ADIAG,Z)
       EPS=1E-16
       NIT=0
       CALL SCOFI1(N,A,ADIAG,U,NIT,EPS)
       CALL ORDR(N,A,ADIAG,U)
       CALL PMAT(N,M,U,PRS)
       CALL FVAL(N,PRS,FV,A)
       CALL OTPT(N,M,A,ADIAG,U,PRS,FV)
       CALL DIMO(N,X,Y,Z,PRS)
       CALL TRMOM(N,M,LVLS,U,X,Y,ADIAG)
       CALL ATAT(N,M,ADIAG,U)
    10 CONTINUE
       STOP
       END



       SUBROUTINE CORDS(N,X,Y,IX,IY)
       DIMENSION X(96),Y(96),IX(96),IY(96),NATM(96)
       EL=1.40
       XT=0.8660254*EL
       YT=0.5*EL
       READ(7,98)(NATM(J),J=1,N)
       DO 10 I=1,N
       ID=NATM(I)
       X(I)=IX(ID)*XT
    10 Y(I)=IY(ID)*YT
    98 FORMAT(40I2)
       RETURN
       END
```

```
      SUBROUTINE MODH(N,A,ADIAG,Z)
      DIMENSION A(30,30),ADIAG(30)
      DIMENSION Z(30)
      READ(7,99)NMOD
      IF(NMOD)18,18,19
   19 WRITE (2,103)
  103 FORMAT (///14H MODIFICATIONS)
      DO 17 K=1,NMOD
      READ(7,100)I,J,X
      WRITE (2,102)I,J,X
  102 FORMAT(I3,2X,I3,3X,F7.3)
      IF (I-J)15,16,14
   16 ADIAG(J)=X
      GO TO 17
   15 A(J,I)=X
      GO TO 17
   14 A(I,J)=X
   17 CONTINUE
   22 READ(7,200) IJ,X
  200 FORMAT(I2,F3.1)
      WRITE(2,201)IJ,X
  201 FORMAT(I3,3X,F4.1)
      IF(IJ)18,18,21
   21 Z(IJ)=X
      GO TO 22
   99 FORMAT(I4)
  100 FORMAT(I2,I2,F6.3)
   18 RETURN
      END



      SUBROUTINE DIMO(N,X,Y,Z,PRS)
      DIMENSION X(96),Y(96),PRS(30,30)
      DIMENSION Z(30)
      EN=N
      SX=0
      SY=0
      DO 18 I=1,N
      P=PRS(I,I)-Z(I)
      SX=SX+P*X(I)
   18 SY=SY+P*Y(I)
      EMU=4.77*SQRT(SX*SX+SY*SY)
      SX=4.77*SX
      SY=4.77*SY
      WRITE(2,109)EMU,SX,SY
  109 FORMAT(/15H DIPOLE MOMENT=,F8.4,2X,5H XMU=,F8.4,2X,5H YMU=,F8.4)
      RETURN
      END
```

```
      SUBROUTINE TRMOM(N,M,LVLS,C,X,Y,ADIAG)
      DIMENSION C(30,30),X(96),Y(96),ADIAG(30)
      DIMENSION LL(64),LH(64),XMIK(64),YMIK(64),TERM(64),OSC(64)
      I=1
      DO 10 J=1,LVLS
      MD=M+1
      MT=M
      IF(J-1)17,17,11
   17 LH(I)=MD
      LL(I)=MT
      GO TO 10
   11 KUPP=J-1
      DO 12 K=1,KUPP
      I=I+1
      LH(I)=M+J
      LL(I)=MT
      I=I+1
      LH(I)=MD
      LL(I)=M-KUPP
      MT=MT-1
   12 MD=MD+1
      I=I+1
      LL(I)=M-KUPP
      LH(I)=M+J
   10 CONTINUE
      MINK=LVLS*LVLS
      DO 18 I=1,MINK
      JI=LL(I)
      JK=LH(I)
      SX=0
      SY=0
      DO 27 L=1,N
      C1=C(L,JI)*C(L,JK)
      SX=SX+C1*X(L)
   27 SY=SY+C1*Y(L)
      XMIK(I)=SX
      YMIK(I)=SY
      SBETA=3.6
      TERM(I)=(ADIAG(JK)-ADIAG(JI))*8067.5*SBETA
   18 OSC(I)=0.0000217*(SX*SX+SY*SY)*TERM(I)
      WRITE(2,200)
      WRITE(2,202)
      DO 19 I=1,MINK
   19 WRITE(2,201)LL(I),LH(I),TERM(I),XMIK(I),YMIK(I),OSC(I)
  201 FORMAT(I4,2X,I4,1X,F10.2,2X,F8.3,2X,F8.3,3X,F8.4)
  200 FORMAT(//19H TRANSITION MOMENTS)
  202 FORMAT(/2X,2H I,4X,2H K,3X,7H ENERGY,5X,5H XINT,5X,5H YINT,5X,5H O
     1SCS)
      RETURN
      END
```

4.7 PROBLEMS

Most of the problems presented below involve sets of calculations in which parameters are varied systematically over prescribed ranges. In all cases they can be accomplished in a single computer run, provided the data is specified adequately and correctly. Several problems are quite long and produce considerable output for inspection; they imply some experimentation with parameter modifications, and are in the nature of small projects that are comparable, in a sense, to laboratory work in experimental sciences.

They are frequently stated briefly, and in outline only and, in certain respects, in an imprecise manner. In this respect they resemble rather brief laboratory notes that indicate a procedure but leave the planning and implementation to the discretion of the investigator. For example, problem 5 is intended to promote a systematic study of the conditions under which certain π-electron excitations can be justifiably interpreted in terms of charge-transfer effects. The problem is stated in a naïve form, that requires some effort on the part of the investigator to formulate a series of calculations, based on previous experience, for some molecule of his own choosing, which will help to illuminate ideas concerning charge-transfer effects in conjugated molecules.

The problems are, therefore, intended to provide a basis for the development of experimental work on the Hückel model by computer methods. It is hoped that they will generate ideas in different fields of applications to conjugated molecules.

1. Compute Hückel solutions for aniline and similar molecules having the same parent benzyl framework (see Section 6). Apply the modification parameter h_7 through the range 0 (0·5) 3·5; within this range $h_7 = 1·0$, 2·0, 3·0 might represent $-NH_2$, $-OH$ and $-F$ substitution respectively.

2. Repeat problem 1 for α- and β-naphthylamine. Observe the alternating polarity of net charges at successive atom positions r, measured by the quantity $(q_r - 1)$.

3. Compute solutions for $-NO_2$, $-COOH$, $-C = O$ and other mesomeric substituent groups in benzene and naphthalene (see Table 4-1), and compare alternating effects produced by electron donating and accepting groups.

4. Consider the form of the MOs involved in 'excitations' from highest occupied to lowest unoccupied energy levels for each molecular species considered in problems 3 and 1.

To what extent is it possible, if at all, to interpret the relevant excitations as charge transfer effects between R and Y where R–Y is the composite π-electron molecule, and Y stands for $-NH_2$, $-OH$, $-F$, $-NO_2$, $-COOH$, $-CO$? (Use the subroutine TRMOM to pick out forbidden 'excitations' and to determine the polarizations and intensities of those that are allowed.)

5. In a series of calculations reduce the magnitudes of resonance integrals β^* connecting R and Y systematically towards zero and note the changes in orbitals, energy levels and transition moments. To what extent are the transitions of the composite system R–Y best described as charge transfer effects or in terms of 'repulsions' between interacting levels of the components R and Y?

(Note:—As a first example examine the case of α-naphthylamine as considered in problem 2. Take $h_N = +0.8$ and vary $\beta^* = k\beta$ through the values $k = 1.0, 0.75, 0.5, 0.25$ and 0.001. Repeat the calculations with $h_N = +1.2$).

6. Take a linear polyene with, say, ten conjugated carbon atoms, and apply the modification $\delta\alpha_r = h_r\beta$ with $h_r = 0, \pm2, \pm4, \pm6$ to the end atom $r = 1$. Observe how a 'surface state' emerges. Next, apply the same changes $\delta\alpha_r$ at both end atoms $r = 1$ and $r = 10$ simultaneously. Two surface states emerge virtually together, as a degenerate pair. What are the nodal characteristics of the corresponding molecular orbitals? Transform these orbitals by taking sum and difference combinations to give one orbital largely localized on atom 1 and the other on atom 10.

7. Compute solutions for naphthalene, applying the modifications $\delta\alpha_u = h_u\beta$ with $h_u = 0, \pm2, \pm4$, and $u = 1, 2$ in turn. Observe the emergence of levels from within the band, the 'localization' of orbitals, and 'migration' of nodes.

Note the relationships

$$\Delta q_s(\delta\alpha_u) = -\Delta q_s(-\delta\alpha_u)$$
$$p_{st}(\delta\alpha_u) = -p_{st}(-\delta\alpha_u) \qquad (s, t \text{ same set})$$
$$= +p_{st}(-\delta\alpha_u) \qquad (s, t \text{ opposite sets})$$
$$F_s(\delta\alpha_u) = +F_s(-\delta\alpha_u)$$

that derive from the properties of 'conjugate' solutions, and, by including the subroutines ATAT, DIMO, and TRMOM confirm the following additional relationships that stem from the same source

$$\pi_{r,s}(+\delta\alpha_u) = +\pi_{r,s}(-\delta\alpha_u)$$
$$\mu_\pi(+\delta\alpha_u) = -\mu_\pi(-\delta\alpha_u)$$
$$f_{i\to k'}(+\delta\alpha_u) = f_{i\to k'}(-\delta\alpha_u)$$

8. Compute solutions for stilbene

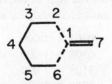

and apply successively the modifications $\beta_{13,14} = k\beta$ with $k = 1\cdot 0$, $0\cdot 75$, $0\cdot 50$, $0\cdot 25$, $0\cdot 001$, 0; these changes might be used to describe, in the Hückel approximation, rotation about the central 13, 14 bond.

Note that the levels change progressively towards those of two 'benzyl' π-electron systems. Transform the MOs of the degenerate levels obtained in the case $k = 0$ to 'benzyl' MOs by taking sum and difference combinations. In what sense is a correct solution obtained, if at all, for the degenerate case?

9. Apply the same changes in k to the two 'single' bond resonance integrals $\beta_{1,13}$, $\beta_{7,14}$ both simultaneously and separately, keeping $\beta_{13,14} = \beta$, and observe how the MOs change towards those of the benzene and styrene π-electron systems.

Calculate the total π-electron energies $\mathscr{E} = 2\sum \epsilon_j$ in each case, and note that the reduction in $\mathscr{E}$ is less for single than for the double bond.

10. Obtain Hückel solutions for a transition-state structure for electrophilic attack similar to that proposed by Mulliken in which four π electrons

are assigned to the pentadienyl framework and two to the pseudo π double bond. Apply a NUCK matrix describing the 'benzyl' framework and introduce the modifications (in MODH)

$$\beta_{1,6} = \beta_{1,2} = 0$$

$$\beta_{1,7} = 2\cdot 5\beta$$

to transform to the proposed configuration.

Now introduce hyperconjugation to bridge the two π-electron segments by increasing $\beta_{1,2}$ and $\beta_{1,6}$ together in systematic steps given by

$$\beta_{1,2} = \beta_{1,6} = k\beta$$

where $k = 0.1(0.1)0.5$.

Observe that hyperconjugation in this representation stabilizes the system by lowering the total π-electron energy.

It should be emphasized that this simple treatment does not do justice to Mulliken's model which incorporated factors representing effective net charges and introduced some degree of self-consistency.

11. Construct the NUCK matrix for $(PN)_4$ using, for example, the labelling shown

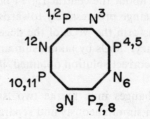

in which

ϕ_r	$(r = 1, 4, 7, 10)$	represent	$P(d_{yz})$ atomic orbitals
ϕ_r	$(r = 2, 5, 8, 11)$	,,	$P(d_{xz})$,, ,,
ϕ_r	$(r = 3, 6, 9, 12)$	,,	$N(p_z)$,, ,,

Each element 1 of the NUCK matrix is converted in the input routine INPT to -1.0 to comply with the sign of the resonance integral β. Alternate $P(d_{xz})$–$N(p_z)$ orbitals mismatch and the appropriate modifications

$$0503 + 1.000$$
$$0806 + 1.000$$
$$1109 + 1.000$$
$$1202 + 1.000$$

must be read in MODH.

Verify that the roots $\pm 2, 0$ of a single 'island' are repeated with four-fold degeneracy.

4.8 REFERENCES

1. A. Streitwieser, *Molecular Orbital Theory for Organic Chemists*, Wiley, New York, 1961.
2. C. A. Coulson, *Proc. Roy. Soc.* (*London*) **A164**, 383 (1938).
3. M. Bradburn, C. A. Coulson, and G. S. Rushbrooke, *Proc. Roy. Soc. Edinburgh* **A62**, 336 (1948).
4. C. A. Coulson and R. Taylor, *Proc. Phys. Soc.* (*London*) **A65**, 815 (1952).
5. H. H. Greenwood, *J. Am. Chem. Soc.* **77**, 2055 (1955).
6. N. Muller, R. S. Mulliken and L. W. Pickett, *J. Am. Chem. Soc.* **76**, 4770 (1954).
7. R. D. Brown, *Australian J. Sci. Res.*, **A2**, 564 (1949).
8. H. H. Greenwood and R. McWeeny, *Advan. Phys. Org. Chem.*, **4**, 73 (1966).
9. G. R. Baldock, *Proc. Cambridge Phil. Soc.* **48**, 457 (1952).
10. E. T. Goodwin, *Proc. Cambridge Phil. Soc.* **35**, 221, 233 (1939).
11. T. B. Grimley, *Proc. Phys. Soc.* (*London*) **72**, 103 (1958).
12. C. A. Coulson, *Proc. Roy. Soc.* (*London*) **A207**, 63 (1951)
13. E. Clar, *Aromatische Kohlenwasserstoffe*, Springer–Verlag, Berlin, 1941.
14. H. B. Klevens and J. R. Platt, *J. Chem. Phys.* **17**, 470 (1949).
15. R. Pariser, *J. Chem. Phys.*, **24**, 250 (1956).
16. D. P. Craig and N. L. Paddock, *Nature*, **181**, 1052
17. D. P. Craig, in *Theoretical Organic Chemistry* (*Kekulé Symposium*), Butterworth, London, 1959, p. 20
18. D. P. Craig, *J. Chem. Soc.* (*London*), 997 (1959).
19. M. J. S. Dewar, E. A. C. Lucken, and M. A. Whitehead, *J. Chem. Soc.* (*London*) 2423 (1960).

5

Reactivity Indices in Molecular Orbital Theory

The reactions of conjugated molecules have been interpreted in MO theory in terms of reactivity indices associated with different models of reaction processes. Such models are generally related, at least notionally, to certain regions of a potential energy curve for the reacting species of the form shown in Figure 5-1. Most indices refer either to region A, which represents the early stages of a reaction, or to region B, which represents a transition state for the reaction. The reactivity indices associated with models describing region A are, or can be, defined in terms of perturbation formulae which describe modifications of π-electron ground states of conjugated molecules, while the most important indices associated with models of region B compute π-electron energy differences between ground states and assumed transition state π-electron configurations. Of the several reactivity indices defined within the framework of MO theory, those associated with the isolated molecule model (Section 5-1A) which refers to region A, and the localization model (Section 5-1B), which refers to region B, only will be considered in detail. Most other indices are defined by similar techniques, though some are less securely based on plausible physical models, and may incorporate parameter adjustments that are questionable on purely theoretical grounds, as indicated by Greenwood and McWeeny.[1]

The reactivity indices of the isolated molecule method are charge densities, free valences, bond orders, and polarizability coefficients, and all are computed by the programs presented earlier. Similarly, the π-electron energies required in applying the localization method, which are particularly tedious to calculate by hand machines, can be obtained from the same programs, from input data describing appropriate parent and residual molecules. Frontier orbital amplitudes (Section 5-4) are, incident-ally, obtained from the same standard computer output, and most other indices can be calculated by the addition of subroutines which use the eigensolutions obtained from the main program.

Computer calculations can be used for studying reactivity problems in two main areas of application. Firstly, in traditional calculations of reactivity indices for the prediction or interpretation of the reactions of conjugated molecules, and secondly to examine the properties of reactivity indices and the models on which they are based. Correlations between numerical values of reactivity indices and observed preferential positions of attack in conjugated molecules have been extensively confirmed, and widely described in the literature, and need not be repeated here. Many references are, for example, given in Part III of Streitwieser's monograph.[2] However, attention has been drawn to discrepancies which can arise in particular cases, and typical examples are considered in Section 5-2, not least because the problems can only be resolved numerically, which implies the use of computer methods. A later section (5-3), shows how computer calculations can assist in providing a working understanding of the nature of relationships between reactivity indices, in a situation where the analyti-cal treatment may be difficult. The examples suggest that, in a wider context, numerical techniques may offer an alternative, more accessible approach than the analytical methods, in the study of theoretical problems, at least until proofs are required. Discrepancies are not, however, con-fined to anomalous predictions of active positions by reactivity indices. Certain reactivity indices which apply to similar regions of the reaction path are, apparently, mutually exclusive on conceptual grounds, though these incompatibilities do not become apparent unless the appropriate numerical calculations are made as shown in Section 5-4. A final section examines the MO interpretation beyond the restrictions imposed by the use of reactivity indices, and suggests a physical interpretation which sup-ports the notion of localized π-complexes as proposed by Olah and his associates.[3-5]

5.1 MOLECULAR ORBITAL THEORIES OF REACTIVITY

In the theory of absolute reaction rates, the rates are determined under conditions of constant temperature and pressure, by the free energy of activation

$$\Delta F_{\ddagger}^{\ddagger} = \Delta H_{\ddagger}^{\ddagger} - T\Delta S_{\ddagger}^{\ddagger}$$

where $\Delta H\ddagger$ and $\Delta S\ddagger$ are enthalpies and entropies of activation. According to the theory the rate constant in solution is given by

$$k = \frac{\mathbf{k}T}{\mathbf{h}} e^{-\Delta F\ddagger/RT}$$

where **k** and **h** are Boltzman's and Planck's constants respectively. The experimental Arrhenuis equation for the rate

$$k = A e^{-\Delta E_a/RT}$$

is expressed in terms of empirical parameters A and ΔE_a, where ΔE_a is the experimental activation energy, which can be related to the thermo-dynamic functions. For example, it can be shown that[6]

$$\Delta H\ddagger = \Delta E_a - RT$$

so that

$$k = \frac{\mathbf{k}T}{\mathbf{h}} e^{\Delta S\ddagger/R} e^{-\Delta H\ddagger/RT}$$

gives

$$k = \frac{e\mathbf{k}T}{\mathbf{h}} e^{\Delta S\ddagger/R} e^{-\Delta E_a/RT}$$

or

$$A \equiv \frac{e\mathbf{k}T}{\mathbf{h}} e^{\Delta S\ddagger/R}$$

It is assumed that, for similar reactions, entropies of activation $\Delta S\ddagger$, involving rearrangements of the σ-bonded framework, are comparable for different positions of attack in the same, or similar, conjugated molecules, and that relative rates are primarily dependent upon the term $\exp(-\Delta E_a/RT)$.

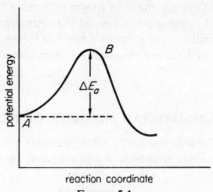

reaction coordinate

FIGURE 5-1

The activation energy ΔE_a may then be subdivided into a term ΔE_π relating to the π electrons only, and a term $\Delta E'$ representing all other contributions to ΔE_a. $\Delta E'$ includes, for example, changes in σ-bond energies, steric and solvent effects, and is assumed to be the same for all positions of attack, so that active positions are ultimately determined by π-electron energy changes only.

A. The isolated molecule method

In their original work on orientation effects in benzene derivatives, Wheland and Pauling[7] assumed that electrophilic and nucleophilic reactions take place preferentially at ring positions of high and low-charge densities respectively, and verified the hypothesis by numerical calculations. In AHs the charge density q_r is unity at all conjugated atoms r, and active positions were then predicted correctly by the charge density q_r' in a polarized molecule, given approximately by

$$q_r' = q_r + \pi_{r,r}\delta\alpha_r \qquad (5\text{-}1)$$

in which $\pi_{r,r}$ is the self-polarizability of atom r, and where the change $\delta\alpha_r$ in coulomb integral at the position r under attack was attributed to the field of the neighbouring charged reagent. Coulson and Longuet-Higgins embodied these, and similar ideas, in a general perturbation theory of π-electron systems[8,9] in which the polarization (5-1) was related to a corresponding approximate change $\delta\mathscr{E}$ in π-electron energy given by

$$\delta\mathscr{E} = q_r\delta\alpha_r + \tfrac{1}{2}\pi_{r,r}\delta\alpha_r^2 \qquad (5\text{-}2)$$

This formula provides a basis for determining active positions by π-electron energy changes expressed in terms of the reactivity indices q_r and $\pi_{r,r}$.

A similar expression relates the free valence F_r to an approximate change $d\mathscr{E}$ in π-electron energy produced by changing resonance integrals β_{rs} and β_{rt} between atom r and its two neighbours s and t; thus

$$\delta\mathscr{E} = 2(\sqrt{3} - F_r)\delta\beta \qquad (5\text{-}3)$$

where $\delta\beta = \delta\beta_{rs} = \delta\beta_{rt}$. Such changes in resonance integrals may be attributed physically to changes in hybridization at the position r under attack associated with incipient σ-bond formation with the incoming reagent. No net charge shifts results from the process (5-3) when applied to AHs, and the free valence is generally assumed to be a reactivity index for radical reactions.

The isolated molecule method applies to the earliest stages A of the reaction process represented in Figure 5-1, where the molecule under attack is exposed to the small changes $\delta\alpha_r$, $\delta\beta$.

B.   The localization method

The method, as originally presented by Wheland,[10] described an activated complex as a resonance hybrid of structures including

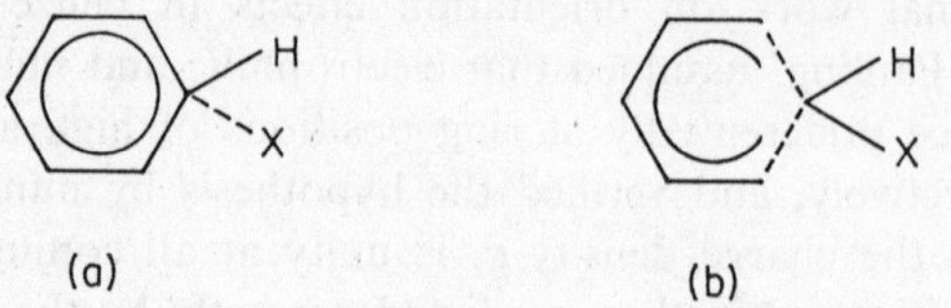

(a)   (b)

FIGURE 5-2

In structure (a) the ionic reagent X lies close, but is not bonded to the atom under attack, and may correspond to a polarized ground state analogous to that described in the isolated molecule approach. In structure (b) the reagent X is joined to the substrate by a roughly tetrahedral σ bond, and the atom r under attack is, in consequence, excluded from conjugation, which is now confined to the corresponding residual molecule (Chapter 4-2). Wheland proposed, in defining the localization method, that structure (b) dominates the resonance hybrid to the virtual exclusion of all other contributions, and that the 'polarization' or localization energy L_r defined as the difference in energy of the π electrons in the unperturbed ground state and of the *same* number of electrons in structure (b) referred to a position r under attack, provides a comparative measure of the ease of attack.

Consider, for simplicity, applications to benzene, where the attacking reagent may be electrophilic (E), nucleophilic (N) or radical (R) in character.

In the three different types of activated complex (E), (R) and (N), two, one, and no π electrons are withdrawn to form the new C–X σ-bond at the position r of attack, and the residual molecules carry in consequence four, five and six π electrons respectively. The energy levels of the penta-dienyl system, $\epsilon = 0$, $\pm\beta$, $\pm\sqrt{3}\beta$ are necessarily the same in each case in the Hückel approximation and the occupancies of levels are shown in Figure 5-4.

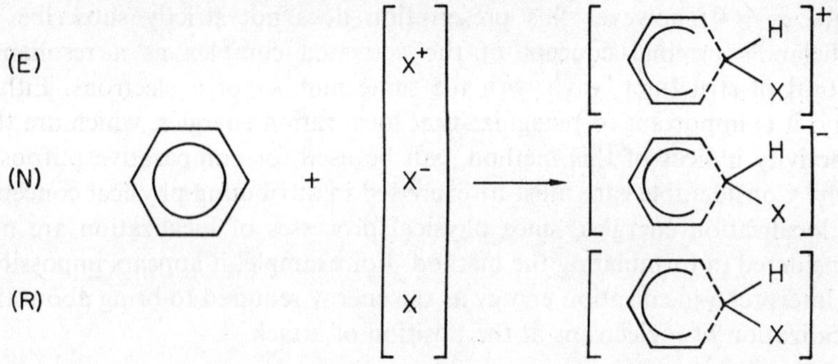

FIGURE 5-3

Wheland assigned an energy $\epsilon = \alpha_r = 0$ to each electron localized at the position r under attack as shown in Figure 5-4, and consequently the localization energies L_r^+, L_r^- and L_r for electrophilic, nucleophilic and radical attack are, for AH, the same. It is an obvious consequence of the charge shifts that the residual molecules are positively charged, negatively charged or neutral in the three cases (E), (N) and (R) respectively, although, in the Hückel approximation, energy levels are not modified accordingly.

The assignment $\epsilon = \alpha_r = 0$ for localized electrons is purely formal; it ensures that localized electrons at all positions r of attack are treated in the same way, so that localization energies may be used for comparative purposes. An alternative prescription that formally omits the energy of localized π electrons, necessarily obtains the same localization energies

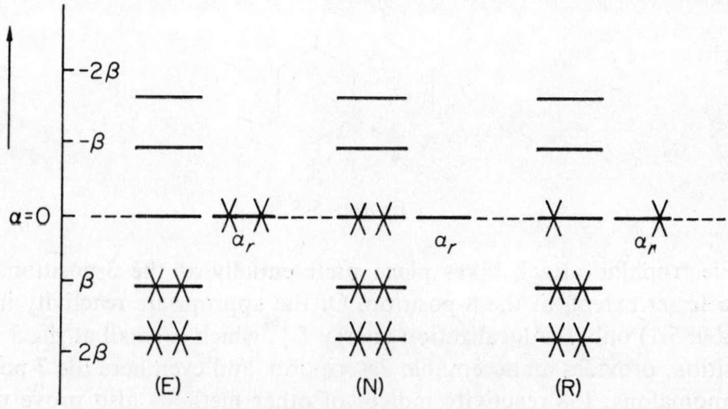

FIGURE 5-4

since $\alpha_r = 0$; however this prescription does not strictly subscribe to Wheland's original concept of the activated complex as a resonance hybrid of structures, each with the same number of π electrons. Either way, it is important to recognize that localization energies, which are the reactivity indices of this method, can be used for comparative purposes only. Considerable care must be exercised in attributing physical concepts to localization energies, since physical processes of localization are not considered in formulating the method. For example, it appears impossible to interpret a localization energy as the energy required to bring about the localization of π electrons at the position of attack.

5.2 REACTIVITY INDICES IN PRACTICE

It is well known that, in general, reactivity indices of both methods predict the same active positions in AHs, and in many derivatives and heterocyclics, though discrepancies can arise in certain non-alternants, heterocyclics and derivatives of AH; localization energies generally predict the active positions correctly, but the reactivity indices q_r and $\pi_{r,r}$ for these systems may fail. Take, for example, the non-alternant fluoranthrene which has been quoted as a typical case illustrating the failure of the isolated molecule method to predict the active position correctly.[2,11,12]

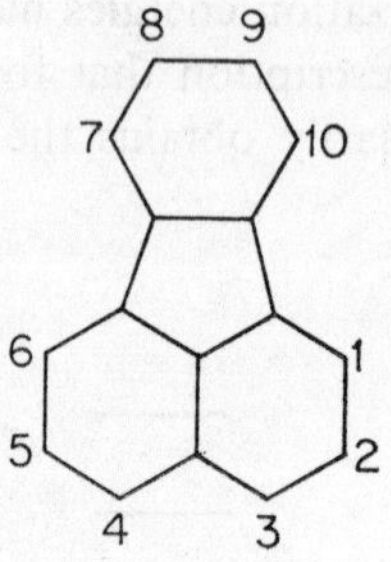

FIGURE 5-5

Electrophilic attack takes place preferentially at the 3-position, and, to a lesser extent, at the 8-position. Of the appropriate reactivity indices (Table 5-1) only the localization energy L_r^+, which is small at the 3 and 8 position, provides an acceptable description, and even here the 7-position is anomalous; the reactivity indices of other methods also prove unsuccessful (Reference 2, p. 347).

Table 5-1

Position	F_r	q_r	$\pi_{r,r}$	L_r^+
1	0·453	0·947	0·440	2·466
2	0·398	1·005	0·400	2·503
3	0·470	0·959	0·462	2·341
7	0·438	0·997	0·427	2·371
8	0·409	1·008	0·410	2·435

Greenwood and McWeeny[1] showed that anomalies which occur in the predictions of active positions by the isolated molecule method may be resolved by considering carefully the implications of the basic formula (5-2). The reactivity indices are slopes and curvatures that contribute terms towards the variation of $\delta\mathscr{E}$ with $\delta\alpha_r$, and cannot be compared independently with experimental observations; only when q_r is the same for all atom positions, as in AHs, is it permissible to consider values of $\pi_{r,r}$ separately. In all other cases both terms in (5-2) must be taken together, and comparisons with experimental observations then necessarily invoke some choice of magnitude for the perturbation parameter $\delta\alpha_r$. When $\delta\alpha_r$ is small, the leading term $q_r\delta\alpha_r$ dominates, and, in fluoranthrene, for example, fails to correlate with experimental results. However when $\delta\alpha_r$ is not small the contributions of both terms balance according to the magnitude of $\delta\alpha_r$; the results obtained from equation (5-2) for fluoranthrene, with $\delta\alpha_r = \beta$ and 2β, suggest (Table 5-2) that position 3 may overtake the remaining positions in 'activity' measured by the approximate energy change $\delta\mathscr{E}$. Similarly the high 'activity' of position 2, predicted by q_2 in Table 5-1 is correspondingly neutralized by the low value of $\pi_{2,2}$, and the values $\delta\mathscr{E}$ $(\delta\alpha_r = 2\beta)$ predict, in fact, the same sequence as L_r^+ in Table 5-1.

Table 5-2

Values of $\delta\mathscr{E}$ given by equation (5-2)

$\delta\alpha_r$	$r =$	1	2	3	7	8
β		1·167	1·205	1·190	1·210	1·213
2β		1·827	1·805	1·883	1·851	1·828

The results of Table 5-2 offer merely a guide to the solution for large values of $\delta\alpha_r$ since the second-order term $\pi_{r,r}\delta\alpha_r^2$ begins to dominate, and fails to predict active positions correctly. Clearly the truncation error of the expansion formula (5-2) prohibits its use in these conditions, and calculations of the 'exact' change $\Delta\mathscr{E}(\delta\alpha_r)$ become imperative. McWeeny and Greenwood[1] have shown that the change $\delta\alpha_r = k\beta$ with $k \sim 2$ to 4 applied in turn to peripheral atoms of fluoranthrene, brings corresponding 'exact' π-electron energy changes $\Delta\mathscr{E}(\delta\alpha_r)$ into agreement with the predictions of localization energies L_r^+ given in Table 5-1. At the same time the 'exact' charge densities q_r' in the corresponding polarized molecules also predict the same sequence of active positions.

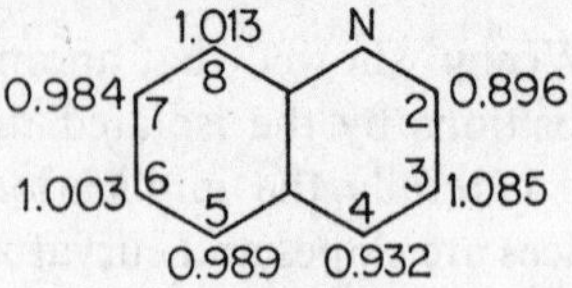

FIGURE 5-6

A second, comparable, example arises in the case of quinolene where the uneven charge distribution stems from the presence of the nitrogen atom. Electrophilic substitution occurs exclusively, and almost equally, at positions 5 and 8, though the charge density distribution indicates $3 > 8 > 6$.[13,14,15] In the parent hydrocarbon $\pi_{1,1} > \pi_{2,2}$ which suggests that the second-order term in (5-2) may, in the heterocyclic, favour the 5 and 8 positions at the expense of the 3 and 6, which carry large charge densities. Numerical calculations[1] confirm this supposition; they show that, as in fluoranthrene, 'exact' changes $\Delta\mathscr{E}(\delta\alpha_r)$ and charge densities $q_r'(\delta\alpha_r)$ in the polarized molecules predict the active positions correctly when $\delta\alpha_r$ is not small. It is an interesting feature of the calculations that large self-polarizabilities $\pi_{r,r}$ at both the 4 position in quinolene and the 1 position in fluoranthrene, do not compensate for the corresponding low-charge densities at these positions, which are correctly predicted to be inactive by 'exact' results $\Delta\mathscr{E}(\delta\alpha_r)$ and $q_r'(\delta\alpha_r)$.

It follows that the reactivity indices q_r and $\pi_{r,r}$ cannot, on theoretical grounds, be correlated with experimental observations, when considered separately, except in the special case of AH for which $q_r = 1$ at all atom positions. For all molecules other than AH, both indices must be considered together, in which case values of $\delta\alpha_r$ must, necessarily, be specified, and, according to the results quoted above, alleged discrepancies between predictions of the isolated molecule method and experimental observations may disappear. Clearly reliable information can be obtained only by computing 'exact' changes $\Delta\mathscr{E}(\delta\alpha_r)$ and $\Delta q_r(\delta\alpha_r)$ by direct solution of the secular equations, over ranges of $\delta\alpha_r$ values, for different positions r of attack, and computer calculations then become imperative.

5.3 THEORETICAL RELATIONSHIPS BETWEEN REACTIVITY INDICES

Many numerical evaluations of reactivity indices belonging to both the isolated molecule and localization methods, predict the same sequence of active positions in AHs. Such correlations were found to be based upon analytical properties of the secular equations,[18] and Baba[16] ultimately established the conditions governing correlations in terms of the integral expansions (3-23, etc.); Fukui and his collaborators[17] summarized these results and provided a comprehensive treatment expressed in terms of the closed forms (3-26, etc.).

All reactivity indices correlate amongst themselves when calculated for AHs, and therefore the reactions of these molecules provide no means of determining the indices associated with given types of reactions. For example, it is not possible to deduce from the reactions of AHs to what extent the free valence is relevant to the case of ionic attack, since it predicts the active positions correctly, like all other indices. In hetero-molecules, however, the indices separate into groups characterized by the dependence upon the hetero system itself, and discrimination becomes possible when comparisons are made with experimental results. These properties are discussed in the following subsections.

A. Reactivity indices in alternant hydrocarbons

Since the localization energies L_r^+, L_r^- and L_r are, by definition, equal when applied to AHs, the relevant correlations are between these indices and those of the isolated molecule method. The following analysis outlines the treatment given by Fukui as it applies to electrophilic reactions.[17]

In the isolated molecule method ionic attack is described by the equation

$$\delta\mathscr{E} = q_r\delta\alpha_r + \tfrac{1}{2}\pi_{r,r}\delta\alpha_r^2 \tag{5-2}$$

where the change $\delta\alpha_r$ in coulomb integral at the position r of attack is attributed to the field of the neighbouring reagent. According to equation (3-20) the self-polarizability appearing in (5-2) can be written in the form

$$\pi_{r,r} = -\frac{1}{\pi}\int y^2 G_r^2 \, \mathrm{d}y \tag{5-4}$$

and, for alternants, q_r is unity. The 'exact' change $\Delta\mathscr{E}(\delta\alpha_r)$ corresponding to $\delta\mathscr{E}$ in equation (5-2) is given by the closed form (3-29)

$$\Delta\mathscr{E} = q_r\delta\alpha_r - \frac{1}{\pi}\int \ln(1 - \delta\alpha_r^2 y^2 G_r^2) \, \mathrm{d}y \tag{5-5}$$

where, again, $q_r = 1$ and G_r is the same term (3-28) appearing in (5-4). Apply now the same change $\delta\alpha_r = \delta\alpha_s = \delta\alpha$ independently at two different conjugated atoms r and s; if

$$\pi_{r,r} \leqslant \pi_{s,s} \tag{5-6}$$

then, since the same terms G_r, G_s arise in (5-4) and (5-5)

$$\Delta\mathscr{E}(\delta\alpha_r) \leqslant \Delta\mathscr{E}(\delta\alpha_s) \tag{5-7}$$

for all $\delta\alpha$. When $|\delta\alpha|$ is large, an energy level associated with a MO largely localized in the region of atom r (or s) emerges from the 'band', and the remaining levels tend towards those of the corresponding residual molecule. This condition associates $\Delta\mathscr{E}(\delta\alpha)$ with a 'localization' energy, and the identification becomes complete when $|\delta\alpha| \to \infty$, as described in Chapter 4, Section 2.

Since $\Delta\mathscr{E}(\delta\alpha_r) \to L_r^+$ (Fukui) as $\delta\alpha_r \to -\infty$ for electrophilic attack, where L_r^+ (Fukui) is the localization energy defined by Fukui's analysis, it follows that if

$$\pi_{r,r} \leqslant \pi_{s,s}$$

which implies

$$\Delta\mathscr{E}(\delta\alpha_r) \leqslant \Delta\mathscr{E}(\delta\alpha_s)$$

for all

$$\delta\alpha_r = \delta\alpha_s = \delta\alpha$$

then as $\delta\alpha \to -\infty$

$$L_r^+ \text{ (Fukui)} \leqslant L_s^+ \text{ (Fukui)} \tag{5-8a}$$

Thus the relationship expressed by the inequalities (5-6, 7 and 8a) ensures the correlation of reactivity indices $\pi_{r,r}$ and L_r^+ (Fukui) describing the electrophilic reactions of AHs. However, the analysis outlined above requires careful consideration of the limiting processes introduced in defining 'localization' energies, especially in the formal treatment of energies assigned to localized electrons, as indicated by McWeeny and Greenwood.[1] In particular, the localization energy introduced in Fukui's analysis cannot be identified with Wheland's definition. In fact, Fukui and his associates[17] proposed a definition in which electrons localized at the position of attack were assigned the limiting value α_r^* of α_r, the coulomb integral at the modified atom r. Then

$$L_r \text{ (Fukui)} = 2\sum_{j=1}^{M}(\epsilon_j^r - \epsilon_j) + (2 - v)(\alpha_r^* - \epsilon_M^r) \tag{5-9}$$

where ϵ_j^r are the levels of the residual molecule RM_r, and ϵ_j are those of the parent AH; $j = M$ identifies the non-bonding MO of the residual molecule, which contains ν electrons ($\nu = 0, 1, 2$ for electrophilic, radical and nucleophilic attack). In the case of electrophilic attack the two electrons localized at the position r of attack are each, therefore, assigned the limiting value $\alpha_r^* = -\infty$. The meaning of the formula (5-9) for electrophilic attack can be deduced from the energy-level diagram (Figure 5-7) associated with modification of the coulomb integral at the rth position, which depicts the emergence of the lowest level ϵ_1, containing the pair of π electrons associated with the MO ψ_1. This MO becomes increasingly localized at the rth position, and all other orbitals and levels change towards those of the residual molecule RM_r. Ultimately, for complete localization ϵ_1 attains the limiting $\alpha_r^* = -\infty$.

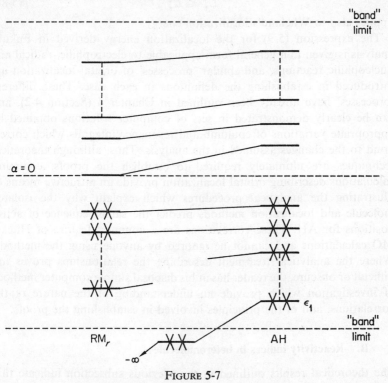

FIGURE 5-7

It will now be recognized that the inequality (5-8a) or

$$L_r^+ \text{ (Fukui)} - L_s^+ \text{ (Fukui)} \leqslant 0$$

implies that the difference $\epsilon_1(r) - \epsilon_1(s)$ in energy of the lowest, emerging levels, becomes zero for complete localization, where each level is associated with the same limiting value of coulomb integral. Thus although the analysis is untidy, in the sense that it requires an unfortunate form of definition (5-9), in which $\alpha_r^* = -\infty$ for electrophilic attack, its meaning is obvious from Figure 5-7.

Wheland's localization energy is not based upon a 'process' of orbital localization, but simply assigns an energy $\alpha = 0$ to each localized π electron. It follows that the two definitions may be related by the formula

$$L_r - L_r \text{ (Fukui)} = (2 - \nu)(\alpha - \alpha_r^*)$$

where the right-hand side is constant for all positions r of attack, in the same or in different AHs. Thus the inequality

$$L_r^+ \leqslant L_s^+ \tag{5-8b}$$

applies also to Wheland's localization energy.

The expression (5-9) for the localization energy derived in Fukui's analysis is given in a general form applicable to electrophilic, radical and nucleophilic reactions, and similar 'processes' of orbital localization are introduced in establishing the definitions in each case. These different 'processes' have already been outlined in Chapter 4 (Section 4-2), and can be clearly demonstrated in sets of computer solutions obtained by appropriate variations of coulomb and resonance integrals which correspond to the changes assumed in the analysis. Thus, although theoretical techniques are ultimately required to establish the proofs, computer calculations describing orbital localization provide an attractive means of illustrating the analytical procedures which explain why the isolated molecule and localization methods predict the same sequence of active positions for AH. These correlations are a common feature of Hückel MO calculations and cannot be ignored by anyone using the methods. Where the analytical treatment describing the relationships proves too difficult or obscure, the reader has at his disposal simple computer methods of investigation which provide an understanding of the nature of the correlations, and of the principles involved in establishing the proofs.

B. Reactivity indices in heteromolecules

The theoretical results outlined in the previous subsection indicate that reactivity indices of the isolated molecule and localization methods inter-correlate amongst themselves when applied to AHs. However, the analytical properties of the reactivity indices differ characteristically in their dependence upon parameters defining the Hückel model, and these differences appear to provide a means of unambiguously associating indices with particular types of reactions of conjugated molecules. Since these properties can be verified numerically by means of computer programs described previously, it will be sufficient, for present purposes, simply to state the relationships and omit the corresponding theoretical proofs.[18]

Assume that the coulomb integral α_u at a site u occupied by a heteroatom is represented by a change $\delta\alpha_u$ from the carbon value α

$$\alpha_u = \alpha + \delta\alpha_u$$

Then the various reactivity indices q_r, $\pi_{r,r}$, F_r, L_r^+, L_r^- and L_r for different positions r of attack change from the values obtained for the parent hydrocarbon, and the relevant properties are those that describe the change in the indices in their dependence upon $\delta\alpha_u$. It turns out that the indices fall into two groups, q_r, L_r^+ and L_r^- which depend upon the sign of $\delta\alpha_u$, and F_r, L_r which are independent of the sign of $\delta\alpha_u$, and these are precisely the groups associated respectively with ionic and free-radical reactions. Experimentally the reactions themselves parallel the changes in indices in their dependence upon electronegativities of directing groups, and, as a result, identification of indices with particular reaction types becomes possible.

The approximate change δq_r in charge density at the rth position due to the change $\delta\alpha_u$ is given by

$$\delta q_r = \pi_{r,u}\delta\alpha_u$$

Whether a prescribed change $\delta\alpha_u$ endows r with (say) electrophilic character depends upon the sign of $\pi_{r,u}$ which alternates in passing from position u; but assume r is chosen for this to be true, so that q_r is increased by δq_r. Then the change $-\delta\alpha_u$ endows nucleophilic character at atom r and q_r decreases by δq_r; this reflects a genuine property discussed in Chapter 3 of the 'exact' change

$$\Delta q_r(+\delta\alpha_u) = -\Delta q_r(-\delta\alpha_u)$$

Analytical properties of the reactivity indices L_r^+ and L_r^- can, likewise, be expressed in their dependence upon the change $\delta\alpha_u$ by the equation[18]

$$\Delta L_r^+(+\delta\alpha_u) = \Delta L_r^-(-\delta\alpha_u)$$

which implies that if the electrophilic index L_r^+ is enhanced by the change $\delta\alpha_u$ then the nucleophilic index L_r^- is enhanced by the same amount by the change $-\delta\alpha_u$. Unlike q_r, neither L_r^+ nor L_r^- is an odd function of $\delta\alpha_u$; in fact

$$\Delta L_r^+(+\delta\alpha_u) \neq -\Delta L_r^+(-\delta\alpha_u)$$

and

$$\Delta L_r^-(+\delta\alpha_u) \neq -\Delta L_r^-(-\delta\alpha_u)$$

and the analysis relates different indices, L_r^+ and L_r^-.

In contrast the change δF_r in free valence is independent of the sign of $\delta\alpha_u$

$$\delta F_r(+\delta\alpha_u) = +\delta F_r(-\delta\alpha_u)$$

and this reflects a corresponding property of the 'exact' change $\Delta F(\delta\alpha_u)$ as described in Chapter 3. The reactivity index L_r possesses a precisely similar property

$$\Delta L_r(+\delta\alpha_u) = +\Delta L_r(-\delta\alpha_u)$$

Thus the reactivity indices of the isolated molecule and localization methods fall into two groups which are distinguished by the analytical dependence upon the sign of a prescribed change $\delta\alpha_u$.

These properties can be illustrated and verified numerically by use of the computer programs by applying equal and opposite changes $\pm\delta\alpha_u$ to the coulomb integral of any atom u of an AH. The numerical values of the changes in the reactivity indices q_r, F_r, L_r^+, L_r^- and L_r for any other atom r of the system, referred to the values obtained for the parent hydrocarbon, will establish the results. A typical set of calculations applied to the AH phenanthrene is presented in problems 3 and 4 at the end of this chapter.

The simplest experimental evidence associating particular types of reactions with particular reactivity indices is found in the reactions of benzene derivatives, as studied originally by Wheland and Pauling.[7] The ortho–para or meta-directing properties are directly related to the electron-donating or attracting properties of a hetero atom or substituent group, and the reactivity indices q_r, L_r^+, L_r^- increase or decrease accordingly to predict the active positions correctly. In contrast F_r and L_r change in the same direction for all directing groups, and radical reactions correspondingly take place at the same positions, as predicted.[7,19] The same general principles are found to be applicable in the reactions of other conjugated molecules.[20]

5.4 ANALYSIS OF MODELS BY COMPUTER METHODS

Computer calculations have been used, so far, for studying apparent discrepancies between the sequences of active positions predicted by reactivity indices of the isolated molecule and localization methods (Section 5-2), and as a means of illustrating analytical relationships between these indices (Section 5-3). We turn now to a different, though related, area of application in which conceptual differences involving other reactivity indices which correlate in predicting the same active positions in conjugated molecules are investigated.

Most indices are defined in terms of coefficients appearing in perturbation formulae, which do not, however, describe the modifications in energy levels and orbitals associated with the perturbations assumed in different models. As a result, it is seldom possible to investigate, in terms of the definitions alone, questions of compatibility amongst indices, and between indices and the physical interpretations claimed for them. On the other hand, the secular equations can always be solved for prescribed values of parameters appearing in perturbation formulae, and some form of discrimination amongst different proposed indices and the associated models may then emerge. As a first example, we discuss incompatibilities between frontier orbital indices and those of the isolated molecule method, both of which are assumed to refer to the early stages of a reaction mechanism.

The frontier electron theory[21] was originally based on the idea that electrophilic reagents would react preferentially with the least bound pair of π electrons of a conjugated molecule which are associated with the highest occupied MO. The magnitude of the density of this orbital at the position r is then adopted as a measure of the susceptibility of atom r towards electrophilic attack. A similar criterion is adopted in the case of nucleophilic reactions, where the frontier orbital is, however, the lowest vacant MO. Frontier orbitals correspond, therefore, to the innermost pair of energy levels, and both are assumed relevant in radical reactions. The density distributions of frontier orbitals in AHs are found to predict, in general, the same active positions as those of the isolated molecule and localization methods, though no analytical relationships have been found which account for this fact.

We shall consider the changes in frontier orbital distributions associated with the formula

$$\delta\mathscr{E} = q_r\delta\alpha_r + \tfrac{1}{2}\pi_{r,r}\delta\alpha_r^2 \qquad (5\text{-}2)$$

which describes, approximately, π-electron energy changes during the early stages of ionic reactions, according to the isolated molecule method, in terms of the reactivity indices q_r and $\pi_{r,r}$. Solutions of the secular equations which yield energy levels and orbitals show, unequivocally, that the frontier orbital density at atom r, which is the assumed position under attack, always diminishes with increase of $|\delta\alpha_r|$ from zero; this applies to any atom r of a conjugated molecule. Thus, although frontier orbital densities in the unperturbed molecule predict active positions correctly, the assumption that this confers a special role in ionic reactions runs contrary to the description given by the isolated molecule method, and to the concept of polarization of π electrons by approaching charged reagents. The frontier orbital hypothesis is, therefore, conceptually incompatible with the isolated molecule model as a criterion for ionic reactions, and can be sustained only if polarization of π-electron systems by incoming charged reagents is rejected. This simple example provides a clear illustration of the need to look beyond numerical correlations amongst indices, and to study the meaning and implications of the models.

Frontier orbitals have been identified with another model[22] embodying the idea of hyperconjugation at the position of attack, which appears to resemble a model of the transition state given by Mulliken and his associates[23,24] Computer calculations again provide the simplest means of identifying the essential features which characterize, in terms of energy levels and orbitals, differences between the two models.

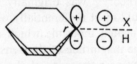

FIGURE 5-8

The π-electron path of conjugation is extended in Fukui's model by the addition of a pseudo π-type orbital ϕ_π^* associated with the attacking reagent X and the hydrogen atom H at the position of attack, as shown in Figure 5-8. The coulomb integral of ϕ_π^* is α^*, and the resonance integral between ϕ_π^* and ϕ_r is β^*. Fukui used perturbation methods, assuming β^* to be small, to derive an approximate expression

$$\delta\mathscr{E} = \sum_{j=1}^{M} \frac{(v_j - v)c_{rj}^2}{\epsilon_j - \alpha^*} \cdot \beta^{*2} + v(\alpha^* - \alpha) \qquad (5\text{-}10a)$$

for the change in π-electron energy due to hyperconjugation. The index j is taken over all MOs of the conjugated molecule under attacT, and v_j is the occupancy; ν is 0, 1 or 2 according as the reagent is electrophilic, radical or nucleophilic. Fukui defines a quantity, the superdelocalizability S_r, by the expression

$$S_r = \sum_{j=1}^{M} \frac{(\nu_j - \nu)c_{rj}^2}{\epsilon_j - \alpha} \tag{5-10b}$$

which differs from the sum in (5-10a) in replacing α^* by the coulomb integral for a carbon atom. It is then argued that since $|\epsilon_j - \alpha|$ is least for the frontier orbitals, with energy levels which lie nearest to α, the summation is dominated by the corresponding term, which for AHs yields

$$S_r' = \frac{2c_{rf}^2}{\lambda_f}$$

where $\lambda_f = |\epsilon_f - \alpha|$, and f denotes a frontier orbital. This equation is assumed to give a theoretical foundation for the frontier orbital hypothesis, based upon the reactivity index S_r.

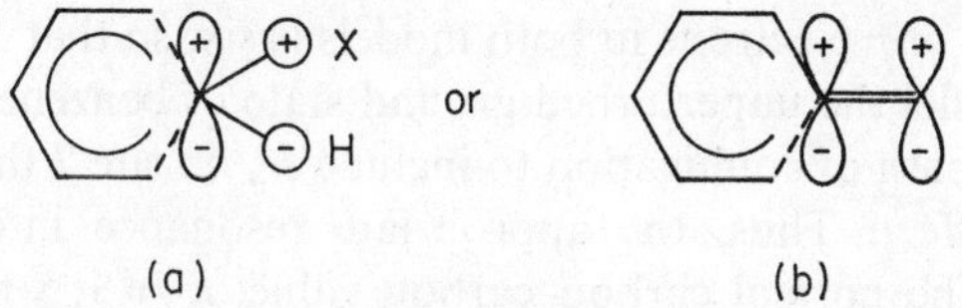

FIGURE 5-9

Mulliken's model is based upon a π-electron transition-state structure that resembles a Wheland hybrid, in which conjugation proper is confined to the residual molecule, and where, for electrophilic attack, two electrons are withdrawn to form a pseudo π double bond. The equivalence of the σ-bond structure (a) and a pseudo π-bond structure (b) is derived from a similar correspondence describing the bonding in ethylene.[25] Hyperconjugation is introduced as a perturbation bridging the residual molecule and the pseudo π bond, both of which, nonetheless, must retain essentially the same character in the resulting π-electron configuration. Mulliken ensured this description by assigning a large resonance integral to the pseudo π bond; certain SCF techniques were also introduced to provide a more reliable account of net charges appearing within the residual molecule as a result of the polarization of the π-electron system. Hyperconjugation is, therefore, superposed as a weak interaction within a π-electron system already polarized, and the model is conceptually compatible with the isolated molecule and localization methods. It is interesting to note that, preceding the introduction of hyperconjugation, the π electrons localized at the position of attack are assigned energies appertaining to the pseudo π double bond, and that this assignment replaces, in principle, the formalism $\epsilon \rightarrow \alpha_r = 0$ introduced in Wheland's method.

A comparison of Fukui's and Mulliken's models of hyperconjugation in the transition state can be obtained from computer calculations based upon the same NUCK matrix, where $r = 1$ is assumed to be the position of electrophilic attack within, for example, the parent AH benzene, and position 7 identifies the site of the pseudo orbital ϕ_π^*.

```
0
1 0
0 1 0
0 0 1 0
0 0 0 1 0
1 0 0 0 1 0
1 0 0 0 0 0
```

The number of π electrons in both models is six, so that $N = 7$, $M = 3$. In Fukui's model the unperturbed ground state of benzene is modified by extending the path of conjugation to include ϕ_π^* on site 7 through a hyperconjugation effect. Thus, the appropriate resonance integral $\beta_{71} = k\beta$ will not take the normal carbon–carbon value, $k = 1$, which appears in the NUCK matrix, but must be modified in subroutine MODH, by for example,

$$0701 - 00 \cdot 125 \qquad \text{(i.e. } \beta_{71} = -\cdot125)$$

In practice, several calculations should be carried out, with $\beta_{71} = 0(-0 \cdot 125) - 0 \cdot 5$ say, so that the changes in levels and orbitals can be recognized more readily. Small values, $\beta_{71} \sim -0 \cdot 125$ might then represent a valid description of hyperconjugation.

Mulliken's model (Figure 5-9) can be based on the same NUCK matrix provided appropriate modifications are made in subroutine MODH. The following modifications

$$0201 \quad 00 \cdot 000$$
$$0601 \quad 00 \cdot 000$$
$$0701 \quad -2 \cdot 500$$

imply that $\beta_{21} = \beta_{61} = 0$ and $\beta_{71} = 2 \cdot 5\beta$, and reproduce Mulliken's model without hyperconjugation, namely the appropriate pentadienyl residual molecule with four π electrons, and a 'strong' pseudo π double bond between atoms 1 and 7. Following this first solution, the effects of hyperconjugation can be introduced by setting $\beta_{21} = \beta_{61} = k\beta$ where $|k\beta|$ is small. The following modifications might be applied in MODH

$$
\begin{array}{ll}
0201 - 00 \cdot 125 & \\
0601 - 00 \cdot 125 & \text{(i.e. } \beta_{21} = \beta_{61} = -0 \cdot 125) \\
0701 - 02 \cdot 500 &
\end{array}
$$

though again, calculations for a set of values, $\beta_{21} = \beta_{61} = 0(-0 \cdot 125) - 0 \cdot 5$ would illustrate the changes more clearly. Note that the same value β_{71} must be reset for each calculation. The parameters used in these calculations are compatible with those used by Mulliken; in particular, the large resonance integral β_{71} largely maintains the double bond character of the pseudo π bond. The results obtained from computer calculations confirm that Mulliken's model is consistent with polarization of the π-electron system, as described in (5-2), which bring towards the position of attack an increased density of the lowest MO, corresponding to the energy level that emerges from below the band.

A few final comments must be added briefly to complete this study of selected models by computer methods. The validity of certain assumptions introduced earlier in deriving definitions of reactivity indices can be checked numerically and preferably by computer methods.

Firstly, the replacement of α^* by α in (5-10a), which makes λ_f the smallest denominator, has neither theoretical nor physical justification; clearly an alternative value could make a different denominator $|\epsilon_j - \alpha^*|$ smallest, and enhance the corresponding term. Furthermore, the assumption that frontier orbital terms dominate the sums (5-10b), following the questionable assignment $\alpha^* \rightarrow \alpha$, for all atoms r of a conjugated molecule, can be shown numerically to be invalid. Individual terms contributing to sums appearing in perturbation formulae can, in all cases where similar assumptions are introduced, be computed, printed, and checked independently.

Next, an assumption that special roles can be associated with the largest (possible dominant) terms of sums appearing in perturbation formulae can always be checked computationally. Such assumptions are by no means confined to the present context, and should always be treated with scepticism. It is valid to quote the familiar case of the self-polarizability

$$
\pi_{r,r} = 4 \sum_{j=1}^{M} \sum_{k=M+1}^{N} \frac{c_{rk}^2 c_{rj}^2}{\epsilon_k - \epsilon_j}
$$

which appears in the formula (5-2) describing ionic attack in the isolated molecule method. The smallest denominator is obtained for the frontier orbitals $j = M$, $k = M + 1$, and the corresponding term is the largest. No special role can be attached to this term, since we know that the frontier orbitals, and all other MOs, except the lowest when $\delta\alpha_r$ is negative, or the highest when $\delta\alpha_r$ is positive, diminish at the position r at which the modification is applied. Indeed, the orbital increasingly localized at r is always associated with large denominators appearing in the sum.

5.5 REACTIVITY INDICES AND LOCALIZED π-COMPLEXES

It was shown in Section 5-2 that application of the perturbation formula

$$\delta\mathscr{E} = q_r\delta\alpha_r + \tfrac{1}{2}\pi_{r,r}\delta\alpha_r^2$$

depends upon a specification of the parameter $\delta\alpha_r$, except in the special case of AHs, for which $q_r = 1$ at all atom positions, and that discrepancies between predictions of the isolated molecule and localization methods disappear when $\delta\alpha_r$ is not small. An estimate of $\delta\alpha_r$ as a function of the distance of separation between the position r of a conjugated molecule and the reagent could, therefore, provide a useful measure of polarization effects, and of the plausibility of the model. Brown[26] has shown that the change $\delta\alpha_r$ due to a unit positive charge X^+ at a distance R from the conjugated atom r can be identified with the interaction term

$$\delta\alpha_r \equiv -\frac{e}{\epsilon R} \tag{5-11}$$

where e is the electronic charge, and ϵ a dielectric constant of the medium. At a distance R $\sim$ 5 A° this corresponds to about 2 electron volts (assuming $\epsilon \sim 1$) so that

$$\delta\alpha_r \sim 2\beta$$

since β, the carbon–carbon resonance integral used in the description of ground-state properties is around 1 eV. This is a large value for $\delta\alpha_r$, comparable in magnitude to those values that remove discrepancies in the isolated molecule description of the reactions of fluoranthrene and quinolene, and it applies to distances R of separation equivalent to 3 or 4 bond lengths. At large distances of separation, the changes $\delta\alpha_s$ at atoms s other than r become virtually equal to $\delta\alpha_r$, since $R_s \simeq R_r$, and it then becomes difficult to understand how active positions can be physically selected at those large distances of separation for which the perturbation formula (5-2) is valid. Selection becomes plausible for close proximities where the active positions are predicted correctly.

It appears that experimental work indicating the existence of σ- and π-complexes as reaction intermediates has some bearing upon this problem, and can make a notable contribution towards a better understanding of the physical significance of reactivity indices. The formation of a σ-complex between an electrophilic reagent and a conjugated molecule involves bond formation at the position of attack by means of two electrons withdrawn from the π-electron system, with the formation of a physically metastable structure which resembles the resonance hybrid introduced

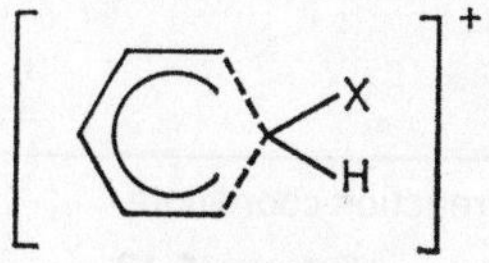

FIGURE 5-10

formally in Wheland's localization method. Dewar's description of a π-complex[27] involves interaction between the electrophilic reagent X^+ and the complete π-electron charge cloud, to form a loose addition complex, represented symbolically in Figure 5-11 in which X^+ is not attached

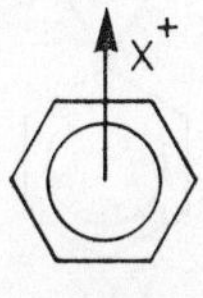

FIGURE 5-11

to a particular atom; many complexes of this kind, involving aromatic molecules and electron acceptors are known. It has been customary to associate π- and σ-complexing with low and high transition states, respectively, depending upon the reactions under consideration; alternatively, σ-complexing may be identified with a local minimum in the region of a main maximum of a potential energy curve, and π-complexing with earlier stages of the reaction path.

Olah and his associates[3-5] deduced a form of potential energy curve from kinetic and orientation effects in the nitration of benzene, and certain derivatives, by nitronium ions derived from $NO_2^+ BF_4^-$ in which properties characterizing σ- and π-complexing were identified. The relevant potential energy curve is represented schematically in Figure 5-12 where the relative heights and depths in the compound barrier must be interpreted qualitatively.

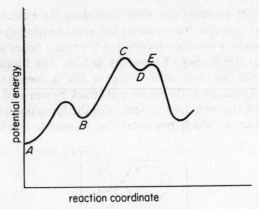

FIGURE 5-12

The energy surface was described by reference to π- and σ-complexing, represented diagrammatically by the structures

The local minimum D in Figure 5-12 is identified with the σ-complex represented in Figure 5-13(b), and the barriers C and E on either side,

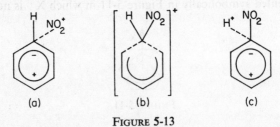

FIGURE 5-13

with transition states through which D is formed and destroyed. Olah was led to deduce, from kinetic and orientation effects, the formation of π-complexes between the aromatic substrate and the $NO_2^+BF_4^-$ ion, and relating to a region B in the potential energy curve on Figure 5-12. Unlike Dewar's model the evidence required a form of localized π-complexing, as described in Figure 5-13(a), with the reagent lying close to the position of attack, and experimental evidence indicated that π-complexing of this kind determined the active positions. The barrier between the regions A and B on the potential energy curve (Figure 5-12) is then associated in Olah's description with dissociation of the (NO_2BF_4)-solvent complex, and the barrier leading to D, with separation of the ion pair; it may also be necessary to add to this hybridization effects involving orbital reorganization at the position of attack.

If localized π-complexing determines the active positions of a conjugated molecule, as proposed by Olah, we may ask to what extent these ideas are compatible with the prediction of active positions by reactivity indices. Localized π-complexing, which assumes that the reagent lies close to the position r of attack, must be associated in the isolated molecule method with large values of $\delta\alpha_r$, the perturbation parameter. These are precisely the conditions under which the validity of the isolated molecule method of prediction can be sustained; otherwise, correct active positions are predicted for small values of $\delta\alpha_r$ only if the sequence of predicted positions does not change with increase in $\delta\alpha_r$. Polarization of the π-electron system in a localized π-complex is also associated in this model with partial localization of a π-electron pair at the position r of attack, and with a lowering $\Delta\mathcal{E}(\delta\alpha_r)$ in π-electron energy which could account, in part, for the observed metastable π-complex associated with the region of the minimum B in the compound barrier. Thus the isolated molecule model, when associated with a large change $\delta\alpha_r$, unambiguously supports Olah's concept of localized π-complexing as the main factor determining active position.

Computer calculations show that for any atom r in an AH, both energy levels and orbitals change rapidly, initially, with changes from zero in $\delta\alpha_r$, and that for values $|\delta\alpha_r|$ greater than about $|3\beta|$, which would apply to the description of localized π-complexes, the levels and orbitals already closely resemble those of the corresponding residual molecule. It could, therefore, be argued that the success of the localization method in predicting active positions in conjugated molecules may be attributed to the close correspondences between π-electron configurations of residual molecules assumed in defining the model used in the localization method, and the π-electron distribution in localized π-complexes.

The various interpretations of the reactions of conjugated molecules provided by different MO models have been the source of considerable controversy. Since many descriptions and models ignore energy level and orbital changes implicit in the definitions of reactivity indices, it is highly desirable that computer calculations should be used to explore the full implications of models, as demonstrated earlier. It is, to some extent, irrelevant whether such investigations extend beyond the range of calculation usually associated with the models. What matters is whether the investigations give coherence, and some degree of conviction to the theoretical interpretation. In the present case, for example, neither the isolated molecule nor the localization model is necessarily bound by formal definition to rigid interpretation, and, when relaxed, both methods provide explanations of orientation effects which support the interpretation deduced from experimental work by Olah, based on localized π-complexes.

5.6 PROBLEMS

1. Compute the charge densities, bond orders, free valences and self-polarizabilities for phenanthrene using the programs of Chapter 2 with

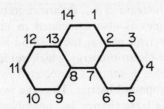

the additional subroutine ATAT. Note the correlation between free valences F_r and self-polarizabilities $\pi_{r,r}$ that represent reactivity indices for free radical and ionic reactions respectively.

2. Compute localization energies L_r for phenanthrene from the total π-electron energies of the residual molecules obtained by excluding successively atoms $r = 1, 3, 4, 5$ and 6 from conjugation.

Note that

$$L_r = \mathscr{E}_\pi \text{ (phenanthrene)} - \mathscr{E}_\pi(\text{RM}_r)$$

correlates inversely with F_r and $\pi_{r,r}$ as obtained in problem 1.

3. Apply equal and opposite modifications, $\delta\alpha_u = \pm\beta$, say, in turn to the atom $u = 1$ in phenanthrene.

Confirm the following relationships

$$\Delta q_r(+\delta\alpha_u) = -\Delta q_r(-\delta\alpha_u)$$

$$\Delta F_r(+\delta\alpha_u) = +\Delta F_r(-\delta\alpha_u)$$

$$\Delta p_{st}(+\delta\alpha_u) = -\Delta p_{st}(-\delta\alpha_u) \quad (r, s \text{ same set})$$

$$= +\Delta p_{st}(-\delta\alpha_u) \quad (r, s \text{ opposite set})$$

for all r (and s), that identify odd or even dependence upon $\delta\alpha_u$.

4. Compute localization energies L_r^+, L_r^- and L_r for peripheral atoms r for equal and opposite modifications $\delta\alpha_u = \pm\beta$ applied to the atom $u = 1$ in phenanthrene.

Denoting by $\Delta L_r(\delta\alpha_u)$ the change in localization energy

$$\Delta L_r(\delta\alpha_u) = L_r(\delta\alpha_u) - L_r(0)$$

between the parent hydrocarbon and the modified molecule for all atoms r, establish the relationships

$$\Delta L_r^+(+\delta\alpha_u) = -\Delta L_r^-(-\delta\alpha_u)$$

$$\Delta L_r(+\delta\alpha_u) = +\Delta L_r(-\delta\alpha_u)$$

and confirm, by comparing the results obtained in problem 1, that ionic and free-radical reactivity indices fall into two groups, according to their dependence upon $\delta\alpha_u$.

5. Suppose the effect of an electrophilic reagent forming a localized π complex at the position r in a conjugated molecule can be represented by a modification $\delta\alpha_r$. Obtain Hückel solutions for $\delta\alpha_r = 0$, β, 3β, 5β and $-\infty$ for positions $r = 1, 3, 4, 5$ and 6 in phenanthrene. Plot the changes in energy levels in each case, and determine whether the localization energies L_r^+ predict the same sequence of active positions as $\Delta\mathscr{E}(\delta\alpha_r)$ for all values of $\delta\alpha_r$. Trace the corresponding changes in MOs obtained for modifications $\delta\alpha_r$ at each atom r. At what stage of modification, if at all, does it become possible to recognize resemblances with those of the corresponding residual molecules?

5.7 REFERENCES

1. H. H. Greenwood and R. McWeeny, *Advan. Phys. Org. Chem.* **4,** 73 (1966).
2. A. Streitwieser Jr., *Molecular Orbital Theory for Organic Chemists*, Wiley, New York, 1961, Part III.
3. G. A. Olah and S. J. Kuhn, *J. Am. Chem. Soc.*, **80,** 6535 (1958).
4. G. A. Olah, S. J. Kuhn and S. H. Flood, *J. Am. Chem. Soc.*, **83,** 4571 (1961).
5. G. A. Olah, S. J. Kuhn, S. H. Flood and J. C. Evans, *J. Am. Chem. Soc.*, **84,** 3684, 3687 (1962).
6. K. J. Laidler, *Chemical Kinetics*, 2nd ed., McGraw–Hill, New York, 1965.
7. G. W. Wheland and L. Pauling, *J. Am. Chem. Soc.*, **57,** 2086 (1935).
8. C. A. Coulson and H. C. Longuet Higgins, *Proc. Roy. Soc. (London)* **A191,** 39 (1947).
9. C. A. Coulson and H. C. Longuet Higgins, *Proc. Roy. Soc. (London)*, **A192,** 16 (1947).
10. G. W. Wheland, *J. Am. Chem. Soc.*, **64,** 900 (1942).
11. R. D. Brown, in *Molecular Orbitals in Chemistry, Physics and Biology*, Academic Press, New York, 1964.
12. K. Fukui, in *Molecular Orbitals in Chemistry, Physics and Biology*, Academic Press, New York, 1964.

13. M. J. S. Dewar and P. M. Maitlis, *J. Chem. Soc.* (*London*), 2521 (1957).
14. D. H. Derbyshire and W. A. Waters, *J. Chem. Soc.* (*London*), 573 (1950).
15. J. H. Ridd, *Phys. Methods Heterocyclic Chem.* **1,** 109 (1963).
16. H. Baba, *Bull. Chem. Soc. Japan*, **30,** 147 (1957).
17. K. Fukui, T. Yonezawa and C. Nagata, *J. Chem. Phys.*, **26,** 831 (1957).
18. H. H. Greenwood, *Trans. Faraday Soc.*, **48,** 585 (1952).
19. D. H. Hey and G. H. Williams, *Discussion Faraday Soc.*, **14,** 216 (1953).
20. E. Fahrenhorst and E. C. Kooyman, *Nature*, **175,** 598 (1955)
21. K. Fukui, T. Yonezawa and H. Shingu, *J. Chem. Phys.*, **20,** 722 (1952).
22. K. Fukui, T. Yonezawa and C. Nagata, *Bull. Chem. Soc. Japan*, **27,** 423 (1954).
23. L. W. Pickett, N. Muller and R. S. Mulliken, *J. Chem. Phys.*, **21,** 1400 (1953).
24. N. Muller, L. W. Pickett and R. S. Mulliken, *J. Am. Chem. Soc.*, **76,** 4770 (1954).
25. G. G. Hall and J. E. Lennard Jones, *Proc. Roy. Soc.* (*London*), **A205,** 357 (1951).
26. R. D. Brown, *Tetrahedron*, **19,** Suppl. 2, 337 (1963).
27. M. J. S. Dewar, *Bull. Soc, Chim. France*, **18,** C71 (1951).

6

The Self Consistent Field Method

The Hückel method has been used exclusively so far, in the interpretation of physical and chemical properties of conjugated molecules. The formulation was based first upon the assumption of $\pi-\sigma$ separation, and then in terms of an 'effective' hamiltonian for π electrons, assuming a π electron moves in the field of a σ-bonded framework and some averaged field due to the remaining π electrons. Further approximations were introduced subsequently to simplify the form of the equations, by neglecting overlap and all resonance integrals other than those referring to neighbouring conjugated atoms. It is proposed now to change the theoretical method by reintroducing π-electron repulsion terms explicitly according to equation (2-2) whilst retaining the remaining approximations of the Hückel method. This modification represents a modest advance on basic Hückel theory which, nevertheless, produces significant theoretical developments that provide a better interpretation of certain physical phenomena, and, usually, improved agreement with experiment. The theoretical developments are associated with the formulation of approximate SCF–CI methods which broaden the scope of theoretical interpretation, which is still based, however, upon the use of determinantal wavefunctions defined in a basis of linear combination of atomic orbitals (LCAO) MOs.

As stated in the Introduction, we are not primarily concerned here with the derivation of the SCF equations, but with computational matters relating to their solution and application. A brief outline of the theoretical foundations of Roothaan's equations appears, therefore, at the beginning of this chapter, and is followed by a description of the simplifications and rearrangements of terms introduced by Pople, Parr and Pariser, in applications to π-electron systems, which lead to a form of equations similar to those obtained in Hückel theory.

The experience gained earlier in solving problems expressed in terms of

Hückel theory, will make it possible to discuss similar SCF problems quite briefly. This chapter, therefore, includes sections which deal briefly with both analytical properties of the non-linear SCF equations, and with selected applications. The most important properties of the SCF equations are those previously described in terms of 'conjugate' solutions in Hückel theory. Precisely similar properties are obtained in SCF theory, of which those associated with the pairing properties for even AH are a special case. Conjugate solutions can be observed in computer calculations based upon systematic variation of parameters, and, as in Hückel theory, they provide a description from which properties of SCF perturbation coefficients can be deduced.

A brief discussion in Section 6-2 on core and electron-repulsion integrals refers essentially to the scheme adopted in the computer programs listed at the end of the chapter, and largely ignores the many investigations devoted to this topic in the literature. In brief, we believe that, on the basis of previous experience, obtained in studying applications of the Hückel method, the reader is well equipped to conduct his own computer-based experiments using the SCF programs. However, it is useful to draw attention to the existence of ' conjugate' solutions, not only for their intrinsic theoretical relevance, but also in the context of planning computer calculations systematically and economically.

It should be emphasized that both this chapter and the following, which introduces CI methods, deal exclusively with closed-shell ground states, in which an even number of π electrons is assigned to MOs in spin-coupled pairs.

The SCF Equations

The hamiltonian operator for a system of n π electrons moving in a σ-bonded framework can be written in the form

$$h_\pi(1, 2, 3, \ldots n) = \sum_{i=1}^{n} h_{\text{core}}(i) + \tfrac{1}{2} \sum_{i>j=1}^{n} \frac{1}{r_{ij}} \qquad (6\text{-}1)$$

where

$$h_{\text{core}}(i) = -\tfrac{1}{2}\nabla^2(i) + V(i) \qquad (6\text{-}2)$$

represents the 'core' hamiltonian, which comprises a kinetic energy term and the potential energy of a π electron in the field of the σ-bonded framework stripped of the π electrons. The summation over the $1/r_{ij}$ repulsion terms is taken between all pairs of π electrons, and the potential energy term V can be written as a sum over terms associated with the N-framework ions

$$V = \sum_{\alpha=1}^{N} V_\alpha \tag{6-3}$$

Antisymmetric wavefunctions Ψ are constructed in determinantal form, from MOs ψ taken as linear combinations

$$\psi = \sum_{r=1}^{N} c_r \phi_r \tag{6-4}$$

of the N atomic $\phi(2p_z)$ orbitals, and the energy of a given π-electron configuration Ψ is then given by

$$\mathscr{E} = \frac{\int \Psi'^* h_\pi(1, 2, 3, \ldots n) \Psi' \, d\tau}{\int \Psi'^* \Psi' \, d\tau} \tag{6-5}$$

The SCF method now seeks to describe a closed-shell ground state by means of a single determinant that minimizes the energy, where the adjustable parameters are the coefficients c_r of the orbitals ϕ_r. A single determinant functions Ψ_0 for $n\ \pi$ electrons is constructed by assigning the electrons formally, in spin-coupled pairs, to M MOs of the form (6-4) so that

$$\Psi_0 = \frac{1}{\sqrt{n!}} \begin{vmatrix} \psi_1(1)\bar{\psi}_1(1)\psi_2(1)\bar{\psi}_2(1) \ldots \psi_M(1)\bar{\psi}_M(1) \\ \psi_1(1)\bar{\psi}_1(2)\psi_2(2)\bar{\psi}_2(2) \ldots \psi_M(2)\bar{\psi}_M(2) \\ \\ \psi_1(n)\bar{\psi}_1(n)\psi_2(n)\bar{\psi}_2(n) \ldots \psi_M(n)\psi_M(n) \end{vmatrix}$$

or, more briefly

$$\Psi_0 = ||\psi_1(1)\bar{\psi}_1(2)\psi_2(3)\bar{\psi}_2(4) \ldots \psi_M(n-1)\bar{\psi}_M(n)|| \tag{6-6}$$

The energy $\mathscr{E}_0$ of the configuration Ψ_0 is then given by

$$\mathscr{E}_0 = 2\sum_{i=1}^{M} I_i + \sum_{i>j=1}^{M} (2J_{ij} - K_{ij}) \tag{6-7}$$

with

$$I_i = \int \psi_i^*(1) h_{core}(1)\psi_i(1) \, d\tau(1)$$

$$J_{ij} = \int \psi_i^*(1)\psi_j^*(2)\frac{1}{r_{12}}\psi_i(1)\psi_j(2) \, d\tau(1, 2) \tag{6-8}$$

$$K_{ij} = \int \psi_i^*(1)\psi_j^*(2)\frac{1}{r_{12}}\psi_i(2)\psi_j(1) \, d\tau(1, 2)$$

The energy E of (6-7) (or (6-5)) is minimized with respect to variation of Ψ_0, where the change

$$\delta E = \sum \delta I_i + \sum (\delta J_{ij} - \tfrac{1}{2}\delta K_{ij}) \tag{6-9}$$

is subject to the orthogonality constraint

$$\int (\delta\psi_i^*)\psi_j \, d\tau + \int \psi_i^*(\delta\psi_j) \, d\tau = 0 \tag{6-10}$$

The constrained minimization problem of (6-9, 10) can be expressed otherwise in terms of a set of non-linear equations, known as Roothaan's equations[1]

$$\sum h_{rs}c_s = \epsilon \sum S_{rs}c_s \qquad (r, s = 1, 2, \ldots N) \tag{6-11}$$

with

$$h_{rs} = f_{rs} + \sum_t \sum_u P_{tu}[(rs|tu) - \tfrac{1}{2}(rt|su)] \tag{6-12}$$

and the familiar auxiliary definitions

$$\psi = \sum c_r \phi_r \tag{6-13}$$

$$S_{rs} = \int \phi_r^* \phi_s \, d\tau \tag{6-14}$$

$$P_{tu} = 2 \sum_{j=1}^{N/2} c_{tj}^* c_{uj} \tag{6-15}$$

where the 'bond orders' P_{tu} are now found for all pairs of conjugated atoms t, u and not neighbours only, and where

$$f_{rs} = \int \phi_r^*(1) h_{\text{core}}(1)\phi_s(1) \, d\tau(1) \tag{6-16}$$

$$(rs|tu) = \int\int \phi_r^*(1)\phi_s(2)\frac{1}{r_{12}}\phi_t(1)\phi_u(2) \, d\tau(1, 2) \tag{6-17}$$

Pople[2] and simultaneously Pariser and Parr[3] introduced simplifications that ultimately reduced Roothaan's equations to a form comparable to those of Hückel theory. Firstly, overlap integrals S_{rs} and all framework 'resonance' integrals f_{rs} between non-neighbouring conjugated atoms were ignored. Then, for consistency, all electron-repulsion integrals that depend upon the overlap of charge clouds were similarly ignored, which leaves as non-zero only the two-suffix terms

$$(rr|ss) = \gamma_{rs} = \int\int \phi_r^*(1)\phi_s^*(2)\frac{1}{r_{12}}\phi_r(1)\phi_s(2) \, d\tau(1, 2) \tag{6-18}$$

The equations then become

$$\sum h_{rs}c_s = \epsilon c_r \qquad (r, s = 1, 2, \ldots N) \qquad (6\text{-}19)$$

with

$$h_{rr} = f_{rr} + \tfrac{1}{2}P_{rr}\gamma_{rr} + \sum_{s \neq r} P_{ss}\gamma_{rs} \qquad (6\text{-}20)$$

and

$$h_{rs} = \beta_{rs} - \tfrac{1}{2}P_{rs}\gamma_{rs} \qquad (6\text{-}21)$$

where $\beta_{rs} \equiv f_{rs}$ is introduced to parallel Hückel nomenclature. The diagonal term f_{rr} which according to (6-2, 3) is given by

$$f_{rr} = \int \phi_r^*(1)[-\tfrac{1}{2}\nabla_1^2 + \sum_{s=1}^N V_s(1)]\phi_r(1)\, d\tau(1)$$

can be partitioned into two terms

$$f_{rr} = \int \phi_r^*(1)[-\tfrac{1}{2}\nabla_1^2 + V_r(1)]\phi_r(1)\, d\tau(1) + \int \phi_r^*(1)[\sum_{s \neq r=1}^N V_s(1)]\phi_r(1)\, d\tau(1)$$

$$= \omega_r + (r|\sum_{s \neq r=1}^N V_s|r) \qquad (6\text{-}22)$$

where ω_r refers to the framework ion r only. Collecting terms in (6-20) gives

$$h_{rr} = \omega_r + \tfrac{1}{2}P_{rr}\gamma_{rr} + \sum_{s \neq r=1}^N [P_{ss}\gamma_{rs} - (r|V_s|r)] \qquad (6\text{-}23)$$

Bearing in mind that the integral $(r|V_s|r)$ represents the potential energy of an electron in the atomic orbital centred on atom r in the field of the screened ion at position s, and that γ_{rs} represents the energy of repulsion between two electrons in orbitals in the same two atoms, Pople suggested that both integrals could be approximated by the inverse distance law, and wrote

$$P_{ss}\gamma_{rs} - (r|V_s|r) = (P_{ss} - Z_s)\frac{1}{R_{rs}} \qquad (6\text{-}24)$$

where R_{rs} is the distance of separation between atoms r and s, and Z_s is the effective, screened, charge at the framework ion s. Since, in fact, the repulsion integrals enter, in part, as adjustable parameters, it has been found more acceptable to write, instead of (6-24)

$$P_{ss}\gamma_{rs} - (r|V_s|r) = (P_{ss} - Z_s)\gamma_{rs} \qquad (6\text{-}25)$$

6A

and the final form of the matrix elements h_{rs} becomes

$$h_{rr} = \omega_r + \tfrac{1}{2}P_{rr}\gamma_{rr} + \sum_{s \neq r}(P_{ss} - Z_s)\gamma_{rs} \qquad (6\text{-}26)$$

$$h_{rs} = \beta_{rs} - \tfrac{1}{2}P_{rs}\gamma_{rs} \qquad (6\text{-}27)$$

which, with the reduced form of Roothaan's equations

$$\sum_s h_{rs}c_s = \epsilon c_r \qquad (r, s = 1, 2, \ldots N) \qquad (6\text{-}19)$$

define the SCF equations for π-electron systems.

6.1 SOLUTION OF SCF EQUATIONS

A formal resemblance between Hückel (2-22) and SCF equations (6-19) can be established by interpreting diagonal elements h_{rr} as coulomb integrals, and h_{rs} when r and s are neighbours as resonance integrals in the SCF scheme, and writing

$$h_{rr} \to \alpha^{SCF}_r$$

$$h_{rs} \to \beta^{SCF}_{rs} = \beta_{rs} - \tfrac{1}{2}P_{rs}\gamma_{rs} \qquad \text{(for } r, s \text{ neighbours)}$$

Similarities between the two methods are particularly relevant in applications to AHs. For example, charge densities P_{ss} in the SCF method are found to be unity and the resultant sum in (6-26) for the remaining atoms $s \neq r$ of the conjugated system is zero, since $Z_s = 1$. The eigenvalues ϵ_j appearing in equation (6-19) are also found to be distributed symmetrically about the value

$$\omega + \tfrac{1}{2}\gamma \qquad (6\text{-}28)$$

which represents the leading terms in (6-26) particularized for carbon atoms, and corresponding eigenvectors are paired, as in the Hückel method (Chapter 3, Section 1). A zero of energy may therefore be fixed, analogous to that ($\alpha = 0$) introduced in the Hückel method, by subtracting the term (6-28) from each diagonal element, so that, in general

$$\alpha^{SCF}_r = \delta\omega_r + \tfrac{1}{2}(P_{rr}\gamma_{rr} - \gamma) + \sum_{s \neq r}(P_{ss} - Z_s)\gamma_{rs} \qquad (6\text{-}29)$$

where $\delta\omega_r = \omega_r - \omega$; then $\alpha^{SCF}_r = 0$ for AHs, as in the Hückel case. McWeeny extended the analogy further, by adopting $h_{rs} = \beta^{SCF}_{rs} = -4\cdot79\,\text{eV}$ for neighbouring atoms in benzene ($P_{rs} = \tfrac{2}{3}$) as the unit of energy equivalent to β in Hückel theory. For AHs, in general, the SCF resonance integral

$$\beta_{rs}^{\text{SCF}} = \beta_{rs} - \tfrac{1}{2}P_{rs}\gamma_{rs}$$

that refers to neighbouring atoms only, then differs marginally from unity, and the remaining off-diagonal elements

$$-\tfrac{1}{2}P_{rs}\gamma_{rs}/4\cdot79 \tag{6-30}$$

are small, or zero if r and s belong to the same set, when $P_{rs} = 0$. The scaling and change of origin (6-28) bring SCF matrices for AHs into close numerical correspondence with those obtained in Hückel theory, and resemblances in computed solutions are explained by these similarities. For AHs the main differences in computed solutions are due to non-zero off-diagonal elements (6-30) between non-neighbouring atoms. Indeed details of differences in the form of solutions which may be of physical or chemical significance, can often be traced back directly to difference in matrix elements. Thus the scaling technique proposed by McWeeny[4] demonstrates features characterizing differences and similarities between the two theoretical methods more clearly and effectively than conventional computational procedures which adopt, for example, electron volts as energy units. It is, therefore, rewarding to explore SCF solutions in terms of McWeeny's transformation even though practical units may ultimately be preferred; this can be achieved in using the SCF computer programs described later, by dividing matrix elements of $\mathbf{h}$ by the scaling factor $4\cdot79$.

However, resemblances between scaled and adjusted SCF and Hückel matrices and solutions, should not disguise the essential structural differences of the equations. Matrix elements in the SCF method, expressed (6-26, 27) in terms of P_{rs} are ultimately dependent (6-15) upon the form of the MOs (6-13). Thus $\mathbf{h}$ cannot be prescribed until the solution coefficients are themselves known, which implies that the equations (6-19) are non-linear in the coefficients. The essence of SCF theory is, in fact, that interactions of electrons, as expressed in the matrix elements of $\mathbf{h}$, depend upon the MOs they occupy, and therefore upon the solutions themselves, defined as that set of MOs which minimizes the energy in the sense of variation theory. An iterative method of solution must, therefore, be used, initiated by a trial bond order matrix $\mathbf{P}'$ (or trial orbitals $\mathbf{c}'$) that permits the construction of a matrix $\mathbf{h}'$ according to the formulae (6-26, 27); this must then be modified in some systematic way to converge towards $\mathbf{h}$. The method used in the SCF computer programs described later employs diagonalization techniques iteratively. A trial matrix $\mathbf{h}'$ is diagonalized to provide new vectors $\mathbf{c}'$ and a new bond-order matrix $\mathbf{P}'$ from which a new matrix $\mathbf{h}'$ is constructed; the process is repeated iteratively until successive matrices $\mathbf{h}'$ agree within prescribed limits, when $\mathbf{h}'$ is equal to $\mathbf{h}$ within the limits specified. The program obtains a trial bond order matrix $\mathbf{P}'$, for initiating the iteration process, by diagonalizing the core framework matrix with diagonal elements given by ω_r and off-diagonal elements by β_{rs} between neighbours, all other elements being zero. For AHs all $\omega_r \to \omega = 0$ are equal and $\beta_{rs} \to \beta$ are the same, and, therefore, regardless of scaling, orbitals and bond-order matrices $\mathbf{P}'$ of the Hückel approximation are obtained for constructing the first $\mathbf{h}'$. Hückel orbitals are not, however, obtained from this process for other types of conjugated molecules.

6.2 SCF CORE AND ELECTRON REPULSION INTEGRALS

It is useful, at this stage, to attempt to give meaning to terms appearing in the matrix elements of $\mathbf{h}$. For example, the term ω_r of equations (6-22, 26) can intuitively be considered in terms of a Schrodinger equation

$$[-\tfrac{1}{2}\nabla^2 + V_r]\phi_r = \omega_r\phi_r$$

associated with the local core atom r, whose eigensolution may be interpreted as describing a valence state of the conjugated atom, with ω_r an atomic valence state ionization potential that can, in principle, be estimated from experimental data. The factor $(P_{ss} - Z_s)$ is the effective net charge at atom s, and $(P_{ss} - Z_s)\gamma_{rs}$ the potential at r due to the net charge at s; the summation in (6-26) therefore represents the total potential at r due to net charges at all other conjugated atoms. Correspondingly, the term $\tfrac{1}{2}P_{rr}\gamma_{rr}$ represents a potential energy of repulsion due to a π-electron charge density P_{rr} at the local core atom r. The integrals β_{rs} are 'resonance' integrals referred to the core framework, and $-\tfrac{1}{2}P_{rs}\gamma_{rs}$ takes the form of an exchange energy.

Electron-repulsion integrals may, in principle, be determined theoretically from the basic formula (6-18) or by semi-empirical methods. For interatomic distances r_{ij} exceeding 2.80 A$^\circ$ Parr[5] and Pariser[3] proposed a formula

$$\gamma_{ij} = \frac{7 \cdot 1975}{r_{ij}}\left[\left(1 + \left\{\frac{D_i - D_j}{2r_{ij}}\right\}^2\right)^{-\frac{1}{2}} + \left(1 + \left\{\frac{D_i + D_j}{2r_{ij}}\right\}^2\right)^{-\frac{1}{2}}\right] \quad (6\text{-}31)$$

based upon the classical interaction of two uniformly charged spheres of diameter

$$D_i = \frac{4 \cdot 597}{Z_i} \times 10^{-8} \, \text{cms} \qquad (6\text{-}32)$$

where Z_i is the effective nuclear charge for the $2p_z$ atomic orbital ϕ_i. For distances less than $2 \cdot 80$ Å an extrapolation formula

$$\gamma_{ij} = \tfrac{1}{2}(\gamma_{ii} + \gamma_{jj}) + ar_{ij} + br_{ij}^2 \qquad (6\text{-}33)$$

was proposed, with parameters a and b determined by fitting the values obtained from (6-31) at $r = 2 \cdot 80$ Å and $3 \cdot 70$ Å. For large distances of separation, when $r_{ij} \gg D_i$, equation (6-31) reduces to the simple form

$$\gamma_{ij} \simeq \frac{14 \cdot 4}{r_{ij}} \qquad (6\text{-}34)$$

The one-centre integral γ_{ii} is sometimes evaluated from ionization potentials I_i and electron affinities A_i according to a relationship

$$\gamma_{ii} = I_i - A_i \qquad (6\text{-}35)$$

proposed by Mulliken.[6]

Resonance integrals β_{rs} may be derived from a proportionality relationship

$$\frac{\beta_{rs}}{\beta} = \frac{S_{rs}}{S}$$

where β and S are 'standard' carbon–carbon resonance and overlap integrals. The standard β is generally obtained by fitting theoretically determined transition energies, predicted by SCF–CI methods, to observed spectroscopic states. More specifically, Parr and Pariser[7] adjusted β and the relevant electron-repulsion integrals γ_{rs} to reproduce exactly the three lowest singlet and the lowest triplet excitations in benzene, and obtained the results

$$\gamma_{11} = 11 \cdot 35; \quad \gamma_{12} = 7 \cdot 19; \quad \gamma_{13} = 5 \cdot 77; \quad \gamma_{14} = 4 \cdot 97; \quad \beta = -2 \cdot 37 \qquad (6\text{-}36)$$

all values being expressed in eV. These are the largest γ_{rs} integrals, and therefore the most important, for any AH, and the proposed scheme that supplements (6-36) with γ_{ij} calculated for larger distances of separation according to (6-34) or (6-31) is adopted for use in the computer programs for SCF–CI calculations described later.

Precise details of the principles invoked in developing the scheme for evaluating SCF integrals as outlined above, have largely been ignored, but can be found in the original accounts given by Parr and Pariser. Various alternative procedures have also been proposed, but the present scheme is chosen for application simply because it is known to work well for AHs.

Certain important precautions must be made in choosing parameters for heteroatoms and conjugating mesomeric groups. Consider first the formal replacement of a conjugated carbon atom by nitrogen as illustrated in pyridine. The nitrogen atom is isoelectronic with carbon in the sense that it contributes one electron to the π-electron system, and the effective core charge Z_N at the nitrogen atom, when the molecule is stripped of π electrons, is unity. In contrast, the nitrogen atom in aniline contributes two electrons to the π-electron system, and the effective nuclear charge is $Z_N = 2$. Similarly, ionization potentials used in assigning values to the

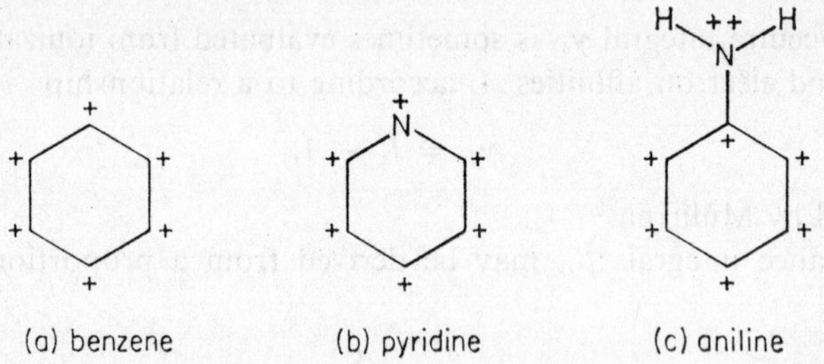

(a) benzene   (b) pyridine   (c) aniline

core charges

FIGURE 6-1

parameters ω_N must refer to singly and doubly ionized valence states of nitrogen in pyridine and aniline respectively. Dewar and Paolini[8] have discussed, in detail, the derivation of SCF parameters for both kinds of nitrogen atoms, as they occur in melamine, and have proposed the set of values (in eV)

Core ion	ω_N	γ_{NN}
N$^+$	14·63	12·27
N^{++}	27·53	14·09

which demonstrates the appreciable difference in values of ω_N obtained for different valence states of conjugated nitrogen atoms.

It should be noted, in passing, that plausible solutions can be obtained by adopting, incorrectly, parameter values for N$^+$ (provided $Z_N = 1$ is also used) in situations referring to doubly ionized nitrogen atoms, as in aniline. The SCF equations are then incorrectly defined, but the results obtained may appear to be acceptable. It is not always the case, especially in earlier SCF studies, that sufficient information has been published to confirm the validity of formulation for doubly ionized conjugated atoms, and these circumstances should be examined with care.

6.3. PROPERTIES OF SCF EQUATIONS

The SCF equations are non-linear in the coefficients c_{rj} that define the solutions, and, therefore, differ significantly from the linear equations of the Hückel method. Yet analytical properties can be deduced,[9] describing parametric variations about solutions for parent even AHs, that are, somewhat surprisingly, virtually identical to those found for the simpler method. These properties will be outlined briefly, and without proofs attached, but their importance should not, as a result, be underestimated. If a prime deficiency of the Hückel method is the explicit neglect of electron-repulsion terms, so that the averaged π-electron field cannot operate to prevent excessive charge concentrations, then the success of the simpler method in describing ground-state properties can be attributed to the agreement in analytical properties between the two methods. For then similar results can always be obtained from the two methods simply by relating magnitudes of parameter variations.[10]

A. Even alternant hydrocarbons

Consider first the SCF description of AHs. It is possible to prove that $P_{rr} = 1$ for all atoms r, and that the levels are paired about the zero of energy $(\omega + \frac{1}{2}\gamma)$, by the methods already described in the context of Hückel equations. Thus, briefly, assume that an eigenvalue ϵ_j referred to the zero $\omega + \frac{1}{2}\gamma$ of energy, and its associated vector

$$\psi_j = \overset{*}{\underset{r}{\sum}}c_{rj}\phi_r + \sum_r c_{rj}\phi_r \qquad (6\text{-}37)$$

are known, where the two summations in (6-37) refer to starred and un-starred atoms respectively. Then $\epsilon'_j = -\epsilon_j$ with an associated vector

$$\psi'_j = \overset{*}{\underset{r}{\sum}}c_{rj}\phi_r - \sum_r c_{rj}\phi_r \qquad (6\text{-}38)$$

is also a solution of the equations, that can be established by inspection, following the technique already described in the case of the Hückel method. Orthogonality conditions and pairing relationships then establish, as before, that

$$P_{rr} = 1$$

and

$$P_{rs} = 0 \qquad (r, s \text{ same set})$$

Pople proved that $P_{rr} = 1$ by an alternative method in which the secular equations were formally solved iteratively, showing that if $P_{rr}^{(n)} = 1$ for all r at the nth stage of iteration, then $P_{rr}^{(n+1)} = 1$; the proof followed inductively by establishing the result for $n = 1$.

It is an interesting feature that the two proofs are theoretically independent. The first describes a property of the SCF equations, and the second an iterative procedure which maintains the condition $P_{rr}^{(n)} = 1$ for all r as $n \rightarrow \infty$, but there is no evidence that $\mathbf{P}^{(n)}$ converges to a unique solution, which can be identified with that implied in the first proof. The fact that many different starting points $\mathbf{P}'^{(n)}$ in which $P_{rr}' \neq 1$ apparently converge towards the same numerical 'solution', in practical applications, inspires confidence in the procedure, but does not ensure theoretical evidence of uniqueness. In fact 'solutions' are progressively corrupted, in practice, by rounding errors as n, the number of iterations, increases.

B. Properties of perturbation coefficients

It is possible to investigate next, properties of changes in solutions due to changes in the parameters ω_u and γ_{uu} at the uth atom position in an AH, and to write

$$\pi_{r,u}^{\text{SCF}} \left(= \frac{\partial P_{rr}}{\partial \omega_u} \right)^{\text{SCF}} \tag{6-39}$$

for the atom–atom polarizability, for example, in a form analogous to that used in Hückel theory. Non-linearity again plays a central role in deducing $\pi_{r,u}^{\text{SCF}}$ since an effective change δP_{rr} must be determined self-consistently; thus the change δh_{uu} in the diagonal element h_{uu} is not simply $\delta\omega_u$ but depends upon the solution $\mathbf{P}(\delta\omega_u)$ itself according to equation (6-26), with a corresponding dependence for all other elements of $\mathbf{h}$.

In spite of the problems associated with non-linearity it is possible to derive analytical properties of the finite changes ΔA ($A \equiv P_{rr}, P_{rs}$ etc.)

$$\Delta A = \sum_{k=1}^{\infty} \frac{1}{k!} \left(\frac{\partial_k A}{\delta \omega_u^k} \right) \partial \omega_u^k \tag{6-40}$$

that depend upon the existence of 'conjugate' solutions analogous to those obtained in the Hückel method. It is, however, useful to express these properties in terms of a modification at the uth position generalized to include variations $\delta\omega_u$ and $\delta\gamma_{uu}$ in both parameters, and to write

$$\delta h'_{uu} = \delta\omega_u + \tfrac{1}{2}P_{uu}\delta\gamma_{uu} \qquad (6\text{-}41)$$

where the prime indicates that part of the total change in the matrix element h_{uu} that contains the parameter changes. Then, denoting by $+$ and $-$ the changes $+\delta h'_{uu}$ and $-\delta h'_{uu}$ the following analytical properties can be proved[9]

(i) $\quad \Delta P_{rr}(+) = -\Delta P_{rr}(-)$

(ii) $\quad P_{rs}(+) = -P_{rs}(-) \qquad (r, s \text{ same set})$

$\qquad\qquad = +P_{rs}(-) \qquad (r, s \text{ opposite set})$

(iii) $\quad \Delta F_r(+) = +\Delta F_r(-)$

(iv) $\quad \epsilon_j(+) = -\epsilon_{N-j+1}(-)$

(v) $\quad c^*_{rj}(+) = c^*_{rN-j+1}(-)$

$\qquad\quad c_{rj}(+) = -c_{rN-j+1}(-)$

These properties express the existence of 'conjugate' solutions corresponding to modifications $\pm\delta h'_{uu}$ and are a generalization of the 'pairing' theorem for SCF equations as established by Pople for AHs. In particular proofs can again be developed in two independent ways,[9] firstly as properties of the SCF equations, and secondly by induction, based upon an iterative method of solution where uniqueness is not established theoretically. These properties (i)–(v) are analogous to those found for the Hückel method, and explain why both methods describe similar changes in π-electron configurations when parameters are modified correspondingly.

The results (i)–(iii) provide descriptions of the properties of the polarizability coefficients of the SCF method; they imply that

$$\frac{\partial^2 P_{rr}}{\partial h'^2_{uu}}, \qquad \frac{\partial F_{rr}}{\partial h'_{uu}}$$

$$\frac{\partial P_{st}}{\partial h'_{uu}} \qquad (s, t \text{ opposite sets})$$

$$\frac{\partial^2 P_{st}}{\partial h'^2_{uu}} \qquad (s, t \text{ same set})$$

are all zero, and that the properties of the charge densities P_{rr}, bond orders P_{st} and free valences F_r with respect to variation of the electro-negativity parameters, are precisely similar in the Hückel and SCF methods.

6.4 VARIATIONS OF PARAMETERS ω AND γ

We have already shown, within the context of Hückel theory, how computational techniques can be organized to demonstrate analytical properties of the method, and similar procedures are applicable in studying the relationships (i)–(v) of the previous section. It is often the case that numerical demonstrations of the existence of such properties, which are readily obtained from computer calculations, are more easily understood and recognized than theoretical methods of analysis, though ultimately such methods are essential in establishing proofs. However, the precise way in which a general modification $\delta h_{uu}'$ may be applied in practice, to study to the relationships (i)–(v) is not obvious, since $\delta h_{uu}'$, which depends upon P_{uu}, cannot be specified a priori, but must be deduced from known solutions of the SCF equations. There is, in fact, a simple way of circumventing this problem which depends upon properties of the net change δh_{uu} as follows.

Suppose that the parameters ω_u and γ_{uu} at the uth position in an even AH are to be varied, and consider the form of the matrix elements

$$h_{uu} = \delta\omega_u + \tfrac{1}{2}(P_{uu}\gamma_{uu} - \gamma) + \sum_{s \neq u}(P_{ss} - Z_s)\gamma_{us}$$

$$h_{rr} = 0 + \tfrac{1}{2}(P_{rr} - 1)\gamma + \sum_{s \neq r}(P_{ss} - Z_s)\gamma_{rs} \qquad (r \neq u)$$

$$h_{st} = \beta - \tfrac{1}{2}P_{st}\gamma_{st} \qquad (6\text{-}42)$$

where all atom positions r other than u are associated with the carbon values ω, γ. Assume now that prescribed modifications $\delta\omega_u^*$, $\delta\gamma_{uu}^*$ generate a solution $\mathbf{P}^*$. Then the same solution $\mathbf{P}^*$ satisfies the SCF equations for any pair of modifications expressed in the general form $\delta\omega_u$, $\delta\gamma_{uu}$ provided

$$\delta\omega_u + \tfrac{1}{2}P_{uu}^*\delta\gamma_{uu} = \delta\omega_u^* + \tfrac{1}{2}P_{uu}^*\delta\gamma_{uu}^* \qquad (6\text{-}43)$$

where the same element $\mathbf{P}_{uu}^*$ of the common solution $\mathbf{P}^*$ appears on both sides of the equation. The proof is obvious, since none of the elements of $\mathbf{h}$ are changed by the substitution (6-43). Suppose we now take the special case in which the framework integral associated with atom u alone is changed by $\delta\omega_u^\dagger$, so that $\delta\gamma_{uu}^\dagger = 0$; then the same solution $\mathbf{P}^*$ holds provided

$$\delta\omega_u^\dagger = \delta\omega_u^* + \tfrac{1}{2}P_{uu}^*\delta\gamma_{uu}^* \qquad (6\text{-}44)$$

A similar condition applies to the special case $\delta\omega_u^{\dagger\dagger} = 0$, $\delta\gamma_{uu}^{\dagger\dagger}$. The relationship (6-44) means that all possible solutions for all modifications can be generated by changing core integrals ω_u only, since all pairs of values $\delta\omega_u^i$, $\delta\gamma_{uu}^i$ generating the same solution can be deduced from

$$\delta\omega_u = \delta\omega_u^i + \tfrac{1}{2}P_{uu}\delta\gamma_{uu}^i \qquad (6\text{-}45)$$

where P_{uu} is an element of the common solution **P**.

These properties clearly promote systematic planning of SCF computations in terms of ω_u variations only, and permit a marked economy of time and effort for solving equations which, in principle, allow independent variation of two parameters ω_u and γ_{uu} per conjugated atom.

The analysis has further implications. Coulomb integrals in Hückel theory are generally associated with relative electronegativities of the conjugated atoms r. Assume that a solution of the SCF equations for a given conjugated molecule is generated by a prescribed set ω_r, γ_{rr} of integrals associated with the $r = 1, 2, \ldots N$ conjugated atoms. Then any alternative pair ω_r', γ_{rr}' of values associated with any rth atom that satisfies

$$\omega_r' + \tfrac{1}{2}P_{rr}\gamma_{rr}' = \omega_r + \tfrac{1}{2}P_{rr}\gamma_{rr} = h_{rr}' \qquad (6\text{-}46)$$

will generate the same solution. The same charge density P_{rr} is associated with the rth site, and P_{rr} varies systematically with h_{rr}' rather than with ω_r or γ_{rr}; it follows that h_{rr}' may usefully be interpreted as the electronegativity parameter in SCF theory. Unlike Hückel theory, the SCF electronegativity h_{rr}' depends, as it should, upon the charge density P_{rr} at atom r, and must change when the same species of atom X, characterized by the same pair ω_X, γ_{XX} of parameters, is sited in different molecules. Thus the electronegativity of a conjugated nitrogen atom $X \equiv N$ is different in pyridine, quinolene, acridine, and so on, just as the associated charge densities P_{NN} are also different. Alternant hydrocarbons provide an exception to this rule, since all carbon atoms are associated with the same pair ω, γ of values, and $P_{rr} = 1$ throughout.

This definition of electronegativity must be revised in systems for which Z_s, the nuclear charge, is not unity, as indicated in the final section of this chapter which discusses SCF π-electron calculations for borazine. In such cases the sums appearing in equations (6-42) which represent the total potential energy term at atom r (or u) due to the net charges at all other conjugated atoms, dominate the matrix elements and largely determine the electronegativities of conjugated atoms. It becomes necessary, therefore, to interpret a concept of electronegativity in terms which relate not only to a given atom but also to its site in the molecule. The electronegativity, which is, in effect, an affinity for distributed charge is, in the

SCF description, dependent not only upon the parameters ω_X, γ_{XX} which characterize an atom, but also upon the total potential field at the site occupied by X, which itself is determined by the π-electron distribution. These properties emerge quite clearly from the study of computer calculations based upon variations of the parameters ω_r, γ_{rr}. They appear in spectacular form in molecules like borazine, which is discussed in the following section, for which Z_s takes values different from unity.

Finally we refer briefly to the calculation of polarizability coefficients in the SCF method, which can be determined in practice from solutions of the SCF equations, using the standard programs, for small changes of the parameters. Numerical values of changes ΔP_{rr}, ΔP_{st}, ΔF_r due to equal and opposite modifications $\pm \delta\omega_u$ (or $\delta h'_{uu}$), which confirm the analytical properties (i)–(v), and the existence of conjugate solutions, can be used to find $\pi_{r,u}$, for example, as the slope at the origin

$$\pi_{r,u} = \frac{\Delta P_{rr}(\delta\omega_u)}{\delta\omega_u} \tag{6-47}$$

The numerical technique (6-47) assumes a small change $\delta\omega_u$ so that higher order terms are negligible, and ΔP_{rr} is effectively linear to the accuracy required with respect to variation of $\delta\omega_u$. In practice this means applying a limit $|\delta\omega_u| \gg 0.25\beta$ to achieve a traditional accuracy to three decimal places in $\pi_{r,u}$. Polarizability coefficients $\pi_{r,u}$ of the SCF method for benzene, naphthalene, anthracene and phenanthrene were originally computed in this way by Greenwood and Hayward,[9] and were subsequently confirmed in a perturbation formulation by Diercksen and McWeeny.[11]

6.5 APPLICATIONS OF SCF–MO METHODS

Changes in the description of ground-state properties of conjugated molecules due to the inclusion of electron-repulsion terms in the SCF method are, generally, less pronounced than the effects produced upon spectroscopic states, where degeneracies are resolved, and the pattern of levels frequently drastically changed when CI techniques are employed. This section discusses two examples only that illustrate the influence of electron-repulsion terms upon the ground-state π-electron configuration.

A. Cyclic polyenes

Ground-state Hückel solutions obtain the same bond orders for all bonds of a cyclic polyene.[12] This result stems, characteristically, from the fact that the Hückel method is a nearest-neighbour approximation that cannot distinguish, for example, between the hexagonal framework of a molecule like (a) cyclo-octadecanonaene $C_{18}H_{18}$ and (b) a ring on which eighteen C atoms are distributed at equally spaced intervals.

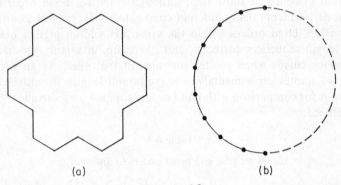

(a)   (b)

FIGURE 6-2

Longuet-Higgins and Salem[13] discussed, by an approximate SCF method, the structure of cyclic polyenes in relation to bond-alternation effects. There is evidence to suggest that in large linear polyenes successive bonds alternate in length, and similar effects were proposed for the cyclic polyenes to explain discrepancies between theoretical predictions and observed spectroscopic states. Longuet-Higgins and Salem simplified the SCF calculations by adopting a model in which all electron-repulsion integrals γ_{rs} referring to non-neighbouring atoms r and s were set equal to zero, and assumed to be absorbed within appropriately chosen nearest-neighbour terms. This nearest neighbour approximation necessarily reproduces equal bond orders for the symmetrical molecule $C_{18}H_{18}$ of Figure 6-2(a). However, the SCF method itself describes properties of π electrons associated with atoms and bonds in their dependence upon the molecular environment by including explicitly terms corresponding to non-neighbouring interactions. As a result, it distinguishes between the six 'outer' and twelve 'inner' bonds of $C_{18}H_{18}$, and calculates two different bond orders, as demonstrated by Murrell and Hinchliffe.[14] Applications to $C_{18}H_{18}$ and other cyclic polyenes therefore, provide useful examples which show that ground-state properties can be examined satisfactorily only if the molecular environment is adequately taken into account, and this, in turn, implies the explicit inclusion of non-neighbour interactions.

It turns out that the numerical values of the two bond orders calculated by the SCF method for $C_{18}H_{18}$ do not agree, in sequence, with the experimentally observed bond lengths if traditional bond order/length relationships are invoked. Murrell and Hinchliffe attempted to account for the apparent discrepancy in terms of σ-bonding effects. There is, however, abundant evidence to show that, although empirical bond order/length curves, derived from many different conjugated AHs, suggest a consistent relationship, bond orders within the same AH seldom predict observed bond length sequences correctly, and, therefore, invariably produce discontinuous curves when plotted one against the other. As an example, Table 6-1 quotes experimentally observed bond lengths in anthracene in column A for comparison with bond orders computed by various theoretical methods.

Table 6-1

Bond lengths and bond orders in anthracene

Bond	A	B	C	D	E
1–2	1·368	0·737	0·842	0·771	0·779
3–4	1·399	0·606	0·632	0·612	0·622
1–1	1·419	0·586	0·449	0·550	0·539
3–3	1·428	0·485	0·548	0·510	0·494
2–3	1·436	0·535	0·418	0·498	0·500

A. X-ray crystallography (Cruickshank)[15]
B. Hückel bond orders (Coulson)[16]
C. SCF bond orders (Pritchard and Sumner)[17]
D. SCF bond orders (Hall)[18]
E. Adjusted Hückel bond orders (Coulson and Golebiewski)[19]

Discrepancies are found between the sequence of Hückel bond orders (column B) and observed bond lengths in anthracene, and these are repeated in naphthalene and, characteristically, in many AHs. Corresponding SCF results obtained by Hall produce a smooth order/length curve in this case, in calculations that are reproduced by the computer programs that follow in Chapter 7, and use Parr–Pariser parameters; the earlier SCF calculations of Pritchard and Sumner use different electron-repulsion integrals which account for differences in computed bond orders. Coulson and Golebiewski employed a Hückel method in which resonance integrals β_{rs} for different bonds r–s were adjusted systematically to preserve a consistent relationship with predicted lengths. Comparable bond order/length curves are obtained for anthracene (and naphthalene) from Coulson and Golebiewski's results and from Hall's calculations, though parameter-adjustment techniques are absent from the latter, which depend upon a self-consistent adjustment of the π-electron distribution only in satisfying the SCF equations.

Most of the evidence, therefore, suggests that the bond order/length relationship is not a universal criterion, and may break down when applied to individual AHs. The situation is somewhat different in Coulson and Golebiewski's technique which imposes a bond order/length relationship in formulating the method. However, this method runs into difficulties when applied to $C_{18}H_{18}$, where it predicts, as a nearest neighbour approximation, equal bond lengths. Returning, therefore, to the original $C_{18}H_{18}$ problem, there is no reason to interpret the SCF solution as an anomalous result which calls for a remedy.

B. Borazine $B_3N_3H_6$

Mention has already been made of the need to ensure that valence state parameters, particularly ω_r and Z_r, are correctly chosen for atoms other than carbon which participate in conjugation. Borazine provides a compact and useful illustration of the method of formulation, and manifests certain analytical properties that describe further interesting features of SCF theory.

Borazine forms a hexagonal structure like benzene, with N and B atoms in alternate positions; boron atoms contribute no electrons, and nitrogen atoms two electrons to the π-electron system, which in the sense of having six π electrons can be compared or contrasted with that of benzene. In fact some flow of π-electron charge from nitrogen to boron atoms must be visualized, and the molecule can, tentatively, be described by an uneven π-electron charge distribution of the form

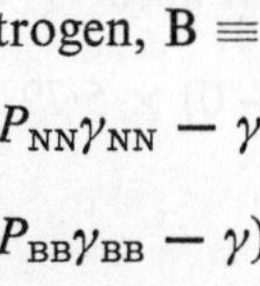

FIGURE 6-3

In the SCF equations (N $\equiv$ nitrogen, B $\equiv$ boron)

$$h_{rr}^N = \delta\omega_N + \tfrac{1}{2}(P_{NN}\gamma_{NN} - \gamma) + \sum_{s \neq r}(P_{ss} - Z_s)\gamma_{rs} \qquad (6\text{-}48)$$

$$h_{rr}^B = \delta\omega_B + \tfrac{1}{2}(P_{BB}\gamma_{BB} - \gamma) + \sum_{s \neq r}(P_{ss} - Z_s)\gamma_{rs} \qquad (6\text{-}49)$$

$$h_{rs} = \beta_{rs} - \tfrac{1}{2}P_{rs}\gamma_{rs} \qquad (6\text{-}50)$$

$Z = 2$ for nitrogen atoms, and zero for boron, and the term, $\omega + \frac{1}{2}\gamma$ which fixes a zero of energy corresponding to a conjugated carbon atom, can be introduced for convenience. The core integrals ω_r refer to doubly ionized nitrogen and neutral boron nuclei, and should be determined correspondingly. Davies[20] obtained solutions for borazine based initially upon the valence-state parameters (in eV)

ω_N	ω_B	γ_{NN}	γ_{BB}
28·85	8·53	15·41	8·21

though these were adjusted to obtain acceptable solutions.

The purpose of the treatment presented below is not to find a plausible solution for borazine with parameter values completely specified, but to illustrate properties of the analytical framework in which solutions may be found. This approach provides not merely for economy, and for a systematic approach to computational studies, but, simultaneously, improved grounds for judging the quality of proposed solutions. In the first place an idealized solution identical with that obtained for benzene is deduced, which provides a reference configuration for describing the nature and properties of solutions obtained by varying parameters describing the system.

Assume, for convenience, that the γ_{rs} and β_{rs} for borazine can be assigned, in the first place, the values (6-36) used in benzene. Then the SCF benzene solution $\mathbf{P}_{BZ}$ with, for example, $P_{rr} = 1$ and $P_{rs} = \frac{2}{3}$ for neighbours, will satisfy the equations for B_3N_3 provided the parameters are adjusted to make $h_{NN} = h_{BB} = 0$. For a N atom the sum in (6-48) takes the value

$$2 \times (1 - 0) \times 7\cdot19 + 2 \times (1 - 2) \times 5\cdot79 + (1 - 0) \times 4\cdot97 = 7\cdot81 \text{ eV}$$
$$(6\text{-}51)$$

so that if

$$\delta\omega_N + \tfrac{1}{2}(\gamma_{NN} - \gamma) = -7\cdot81 \qquad (6\text{-}52)$$

then $h_{NN} = 0$. Similarly the sum in (6-49) takes the value

$$2 \times (1 - 2) \times 7\cdot19 + 2 \times (1 - 0) \times 5\cdot79 + (1 - 2) \times 4\cdot97 = -7\cdot81 \text{ eV}$$
$$(6\text{-}53)$$

so that if

$$\delta\omega_B + \tfrac{1}{2}(\gamma_{BB} - \gamma) = +7\cdot81 \text{ eV} \qquad (6\text{-}54)$$

then $h_{BB} = 0$. For this 'idealized' situation, therefore, the benzene matrix $\mathbf{P}_{BZ}$ is the solution of the SCF equations for borazine and, clearly, the same solution is obtained by the particular choice of parameters

$$\gamma_{NN} = \gamma_{BB} = \gamma; \qquad \delta\omega_B = -\delta\omega_N = 7{\cdot}81 \text{ eV.}$$

In fact, 'conjugate' solutions referred to $\mathbf{P}_{BZ}$ can be generated most simply by variation of the parameters ω_B and ω_N only, though the same solutions apply also to appropriately chosen pairs of values $(\omega'_B, \gamma'_{BB})$ and $(\omega'_N, \gamma'_{NN})$ as demonstrated earlier for AHs. The properties describing 'conjugate' solutions can be expressed in terms of the variation

$$\delta\omega = (\omega_B - \omega_N) - 15{\cdot}62$$

by the following relationships

$$\Delta P_{rr}(+\delta\omega) = -\Delta P_{rr}(-\delta\omega)$$
$$P_{rs}(+\delta\omega) = +P_{rs}(-\delta\omega) \qquad (r, s \text{ opposite sets})$$
$$= -P_{rs}(-\delta\omega) \qquad (r, s \text{ same set})$$
$$F_r(+\delta\omega) = +F_r(-\delta\omega)$$

These formulae are associated with corresponding relationships between energy levels and orbitals for 'conjugate' solutions which are analogous to those obtained for AHs, where the reference configuration $\mathbf{P}_{BZ}$ is obtained when $\delta\omega = 0$.

The analogy can, in fact, be carried further. Atom–atom polarizabilities for benzene, and other coefficients relating to the use of perturbation methods could, in principle, be applied with equal validity to the reference configuration $\mathbf{P}_{BZ}$ in borazine. It is unlikely that such a procedure could be of significant practical value, although changes of opposite signs in core integrals $\delta\omega$ for nitrogen and boron atoms could be considered to yield approximate solutions for borazine.

A set of numerical results for borazine, obtained from the computer programs listed at the end of Chapter 7, are quoted in Table 6-2. These apply to changes $\delta\omega = \pm 1$ eV which illustrate some of the properties expressed in the formulae referred to above.

The form of the diagonal elements given in (6-48) and (6-49) again raise interesting aspects of the SCF description of relative electronegativities. Clearly, a difference $\omega_B - \omega_N = 15{\cdot}62$ eV in framework parameters gives a uniform charge density distribution with $P_{NN} = P_{BB} = 1$, and $P_{NB} = \frac{2}{3}$, corresponding to $h_{NN} = h_{BB} = 0$. The contributions $(P_{ss} - Z_s)\gamma_{rs}$ to the

Table 6-2

Borazine SCF solutions

ω_B	7·81	7·81	6·81	7·81	8·81
ω_N	−7·81	−6·81	−7·81	−8·81	−7·81
$\delta\omega$	0	−1	−1	+1	+1
ϵ_6	−8·705	−8·210	−9·210	−9·210	−8·21
$\epsilon_4 = \epsilon_5$	−5·595	−5·125	−6·125	−6·125	−5·125
$\epsilon_2 = \epsilon_3$	5·595	6·125	5·125	5·126	6·125
ϵ_1	8·705	9·211	8·211	8·211	9·211
P_{NN}	1·0	0·877	0·877	1·123	1·123
P_{BB}	1·0	1·123	1·123	0·877	0·877
P_{BN}	0·667	0·662	0·662	0·662	0·662

diagonal elements, as detailed in (6-51, 53), show that π-electron charge released from nitrogen to boron atoms produces, in reverse, strong negative potentials at nitrogen atom positions which counterbalance the strong local positive potentials at the same positions of the framework; and the effect is reversed at boron atoms. In consequence, differences $\omega_B - \omega_N$ in framework parameters must, with the parameters used above, exceed around 15 eV for the nitrogen atoms to retain charge densities exceeding unity. Clearly, then, relative electronegativities cannot, even approximately, be related directly to framework parameters ω_r, but depend upon both the local field at r and the π-electron charge distribution throughout the rest of the molecule. In particular the electronegativity of a conjugated atom X depends upon its environment in the molecule, and cannot be regarded as an invariant property of the atom.

6.6 NOTE ON PROGRAMS AND PROBLEMS

Programs for computing SCF energy levels and orbitals and other quantities describing the ground-state configuration in the SCF approximation are given at the end of Chapter 7. These programs are designed to form a suite of programs that include CI calculations as a sequel to the SCF solutions. It is desirable, in planning the computations economically, to apply the SCF eigensolution directly to construct matrices of the CI problem; data and parameter cards defining the SCF–CI project are also advantageously introduced within the same procedure. The SCF–CI project is, therefore, best treated as a single problem and for this reason specifications of data for solving particular problems are also postponed.

6.7   REFERENCES

1. C. C. J. Roothaan, *Rev. Mod. Phys.*, **23**, 69 (1951).
2. J. A. Pople, *Trans. Faraday Soc.*, **49**, 1375 (1953).
3. R. Pariser and R. G. Parr, *J. Chem. Phys.*, **21**, 466, 767 (1953).
4. R. McWeeny and T. E. Peacock, *Proc. Phys. Soc. (London)*, **A70**, 41 (1957).
5. R. G. Parr, *J. Chem. Phys.*, **20**, 1499 (1952).
6. R. S. Mulliken, *J. Chim. Phys.*, **46**, 497, 695 (1949).
7. R. G. Parr, *Quantum Theory of Molecular Electronic Structure*, Benjamin New York, 1964.
8. M. J. S. Dewar and L. Paolini, *Trans. Faraday Soc.*, **53**, 261 (1957).
9. H. H. Greenwood and T. H. J. Hayward, *Mol. Phys.* 3, 495 (1960).
10. R. McWeeny, *Proc. Roy. Soc. (London)*, **A237**, 355 (1956).
11. G. Diercksen and R. McWeeny, *J. Chem. Phys.*, **44**, 3554 (1966).
12. C. A. Coulson and A. Streitwieser, *Dictionary of π-Electron Calculations*, Pergamon, New York, 1965.
13. H. C. Longuet Higgins and L. Salem, *Proc. Roy. Soc. (London)*, **A257**, 445 (1960).
14. J. Murrell and A. Hinchliffe, *Trans Faraday Soc.*, **62**, 2011 (1966).
15. D. W. J. Cruickshank and R. A. Sparks, *Proc. Roy. Soc. (London)*, **A258**, 270 (1960).
16. C. A. Coulson, *Proc. Roy. Soc. (London)*, **A207**, 91 (1951)
17. H. O. Pritchard and F. H. Sumner, *Proc. Roy. Soc. (London)*, **A226**, 138 (1954).
18. G. G. Hall, *Trans. Faraday Soc.*, **53**, 573 (1957).
19. C. A. Coulson and A Golebiewski, *Proc. Phys. Soc. (London)*, **78**, 1310 (1961).
20. D. W. Davies, *Trans. Faraday Soc.*, **56**, 1713 (1960).

7

Configuration Interaction and Excited States

The calculation of spectroscopic states of AHs by CI methods resolves degeneracies which arise in the Hückel and SCF descriptions. These degeneracies stem from the pairing properties of levels and orbitals as described in previous chapters, and their resolution in the CI method invariably produces descriptions of spectroscopic states broadly in agreement with experiment, and markedly different from those predicted by the simpler theoretical methods.

This chapter contains, in the first part, a brief outline of the techniques involved in formulating the CI problem for π-electron systems. Formulae for the matrix elements of the π-electron hamiltonian operator defined in a basis of single orbital replacement configurations derived from a SCF ground state are quoted. These formulae form the basis for computer programs which are listed at the end of the chapter and solve the π-electron CI problem by calculating singlet and triplet excitation energies and states. However, no attempt is made to reproduce the derivation of the equations and formulae, and, as in previous chapters, it is assumed that the relevant analysis will be sought in other texts.

The CI programs are linked to follow the set of SCF subroutines mentioned in the previous chapter to form a SCF–CI package, which is particularly easy to use in practice, since the form of input data is virtually identical with that already introduced in applying programs of the Hückel method. The main new problem encountered in applying the programs lies in interpreting the results. The first part of the program output is straightforward, since it presents the solution of the SCF equations and prints SCF energy levels and orbitals, charge densities, bond orders, free valences and dipole moments. However, the output for the second, CI stage is more complicated, and is, therefore, discussed in some detail, with special reference to the results obtained in practice, for naphthalene, and, by way of comparison and contrast, for quinolene.

Numerical values of matrix elements for naphthalene, as obtained from the computer programs, are given in Table 7-2 and a solution of the corresponding CI problem is reproduced in Table 7-3. These results, and all those that follow in the rest of the chapter, refer to the description of excited singlet states only. However, corresponding results for excited triplet states are obtained and printed automatically by the programs. Normally the program prints the solution of the CI problem in terms of state energies E_j and wavefunctions Θ_j with transition moments and polarizations, but it is useful to be able to refer back to the CI matrix itself (Tables 7-2 and 7-5) to illustrate the nature of the problem, and to gain an understanding of the properties of solutions obtained from the method.

Following the treatment of naphthalene, a similar analysis is applied to the molecule quinolene. This examines the effect of heteroatom substitution in removing degeneracies associated with the parent hydrocarbon in the SCF approximation, and its influence on the corresponding CI solution. A later section describes the computation of transition moments for singlet excitations with special reference again to naphthalene and quinolene, and shows how the results obtained can be correlated with coefficients appearing in the wavefunctions describing excited states. A final section shows how to prepare input data for a few chosen π-electron calculations.

7.1 THE CONFIGURATION INTERACTION METHOD

The method describes a state Θ of a system of $n\,\pi$ electrons as a linear combination of configurations Ψ_r each of which represents an assignment of the $n\,\pi$ electrons to an available set of MOs ψ, so that

$$\Theta = \sum C_r \Psi_r \qquad (7\text{-}1)$$

The configurations Ψ_r chosen for inclusion in the expansion form (7-1) depend, in practice, essentially upon the nature of the problem to be solved. Singlet and triplet excited states of conjugated molecules, which are discussed in this chapter, can be approximated effectively by choosing appropriate configurations $\Psi(i \rightarrow k')$ corresponding to transfers of a π electron from an occupied MO ψ_i of the ground-state configuration Ψ_0 to an unoccupied, or virtual orbital $\psi_{k'}$. Self-consistent field MOs ψ_j calculated by the methods described in Chapter 6, will be used exclusively for constructing configurations $\Psi'(i \to k')$, where (equation 6-6)

$$\Psi_0 = ||\psi_1(1)\bar{\psi}_1(2) \ldots \psi_i(\mu)\bar{\psi}_i(\mu + 1) \ldots \psi_M(n - 1)\psi_M(n)|| \quad (7\text{-}2)$$

represents the SCF ground-state configuration. For an even AH M, the number of occupied orbitals will be equal to $N/2$, where N is both the number of molecular and atomic orbitals, and $n = N$. Replacement of the occupied orbital ψ_i by a virtual orbital $\psi_{k'}$ used in constructing singlet configurations $\Psi'(i \to k')$ can be achieved in two ways, depending upon spin assignments, and $\Psi'(i \to k')$ is, therefore, described by two Slater determinants

$$\Psi'_{i \to k'} = \frac{1}{\sqrt{2}} [||\psi_1(1)\bar{\psi}_1(2) \ldots \psi_i(\mu)\bar{\psi}_{k'}(\mu + 1) \ldots \psi_M(n - 1)\psi_M(n)||$$
$$- ||\psi_1(1)\bar{\psi}_1(2) \ldots \psi_{k'}(\mu)\bar{\psi}_i(\mu + 1) \ldots \psi_M(n - 1)\psi_M(n)||] \quad (7\text{-}3)$$

Pi-electron states are determined in the CI method by minimizing the expression for the energy

$$E = \frac{\int \Theta^* h_\pi(1, 2, \ldots n)\Theta \, d\tau}{\int \Theta^* \Theta \, d\tau} \quad (7\text{-}4)$$

with respect to variation of the coefficients C_r of the formula (7-1), where h_π is the operator (6-1)

$$h_\pi(1, 2, 3, \ldots n) = \sum_{i=1}^{n} h_{\text{core}}(i) + \tfrac{1}{2} \sum_{i>j=1}^{n} \frac{1}{r_{ij}} \quad (7\text{-}5)$$

Minimization of the form (7-4) corresponds to solving the eigenvalue problem

$$h_\pi\Theta = E\Theta \quad (7\text{-}6)$$

where the matrix elements of $\mathbf{h}$ are given, in the basis of configurations Ψ_r by

$$h_{IJ} = \int \Psi_I^* h_\pi(1, 2, \ldots n)\Psi_J \, d\tau \quad (7\text{-}7)$$

Simplifications in the form of $\mathbf{h}$ result from a choice of SCF MOs for constructing configurations Ψ_r. In the first place, matrix elements h_{IJ} connecting configurations $\Psi'(i \to k')$ (7-3) resulting from single replacements, and the ground state Ψ_0 are zero. The eigenvalue problem can, therefore, be reduced by omitting Ψ_0 from the expansion form (7-1), provided the lowest excited states are expressed as linear combinations of single-orbital replacement configurations only. This restriction has an additional advantage in that all single replacement configurations are easily identified and can, for a large range of molecules, be included automatically in the computer calculations. Thus, for AHs, when the parameter named LVLS in the computer program is given a value equal to the number of unoccupied energy levels, all possible single-replacement configurations, which are MINK $= (LVLS)^2$ in number, are automatically included in the CI expansion (7-1). The problem is less well defined when higher order replacements are taken into consideration, since the configurations must be selected intuitively from the large number that arise in practice. This does not imply that higher order 'excitations' involving two or more replacements are unimportant, especially in modifying predictions in regions of higher excitation energies, but that restriction to single replacements provides a well-defined theoretical framework, which works effectively in practice, and can be accommodated satisfactorily on a medium-sized computer.

The matrix elements h_{IJ} for single replacements based on SCF orbitals take a simple form that can be derived by the methods described by Condon and Shortley and others.[2,3,4,5] It is convenient to adopt E_0, the energy of the SCF ground state as the zero of energy by effectively subtracting E_0 from all diagonal terms of the CI matrix $\mathbf{h}$, so that[5,6]

$$\langle i \to k' | h_\pi | i \to k' \rangle = \Delta E_{ik'} \pm (ik' | k'i) \qquad (7\text{-}8)$$

with
$$\Delta E_{ik'} = \epsilon_{k'} - \epsilon_i - [(ik' | ik') - (ik' | k'i)] \qquad (7\text{-}9)$$

where the $+$ and $-$ signs refer to singlet and triplet configurations respectively. The off-diagonal elements are given by[5,6]

$$\langle i \to k' | h_\pi | j \to l' \rangle = -[(il' | jk') - (il' | k'j)] \pm (il' | k'j) \qquad (7\text{-}10)$$

with the same rule for signs. In the formula (7-9) ϵ_j are SCF orbital energies, and $(\lambda\mu | \nu\rho)$ are electron-repulsion integrals expressed in terms of SCF MOs $\psi_\lambda, \psi_\mu, \ldots$ etc. These integrals are obtained by transformation from the given basis of atomic orbitals, that simplifies, in the overlap approximation, to the form

$$(\lambda\mu | \nu\rho) = \sum_{rs} c_{r\lambda} c_{r\mu} c_{s\nu} c_{s\rho} \gamma_{rs} \qquad (7\text{-}11)$$

where γ_{rs} are the electron-repulsion integrals (6-18) referring to atomic orbitals ϕ_r, ϕ_s.

The relevant matrix elements (7-8) and (7-10) are calculated in a computer program described later, and the CI matrix $\mathbf{h}$ is diagonalized to give state energies E_j and wavefunctions Θ_j by the JACOBI routine used previously. Implementation of the computer programs represents a fairly straightforward extension of the SCF programs.

A. Alternant hydrocarbons

The influence of CI in determining excited states in conjugated molecules can be illustrated most simply by the reference to the results obtained in some particular application.

Consider the case of naphthalene, which, by the methods described in the previous chapter, yields the SCF orbital energies ϵ_j quoted, in eV, in Table 7-1.

Table 7-1

SCF orbital energies for naphthalene (eV)

j	1 10	2 9	3 8	4 7	5 6
ϵ_j	$\mp 9\cdot46$	$\mp 7\cdot57$	$\mp 6\cdot56$	$\mp 5\cdot40$	$\mp 4\cdot37$

The SCF orbital energies ϵ_j are distributed symmetrically in pairs about the chosen zero of energy $\omega + \frac{1}{2}\gamma$ with occupied bound levels, ϵ_j $(j = 1, 2, \ldots 5)$ negative, and unoccupied levels ϵ_j $(j = 6, 7, \ldots 10)$ positive.

The CI matrix constructed from the nine possible single replacements amongst SCF MOs describing the six innermost levels, $i = 3, 4, 5$ and $k' = 6, 7, 8$ is given in Table 7-2.

The computer program described later defines a data record LVLS essentially similar to that used in the Hückel programs which selects automatically all possible single replacement configurations from amongst the $2 \times$ LVLS innermost SCF MOs; these are (LVLS)^2 in number. For example, the maximum value of LVLS in naphthalene is 5, corresponding to five occupied levels $i = 1$–5 and five unoccupied $k' = 6'$–$10'$, and this choice selects all 25 single replacement configurations that can be constructed for the molecule. The matrix elements given in Table 7-2 were obtained from the computer program by setting LVLS = 3, which selects $i = 3, 4, 5$ and $k' = 6', 7', 8'$, giving nine replacement configurations $\Psi(i \to k')$, and the accompanying diagram illustrates the origin of the first four replacements $(i = 4, 5; k' = 6', 7')$.

Diagonal elements of the CI matrix which are computed according to the formula (7-8) represent configuration energies, and, therefore, energies of excitation in the SCF description, measured relative to the ground-state energy E_0. It will be observed that the configuration energies are not obtained as differences of the orbital energies ϵ_j, given in Table 7-1, but

Table 7-2

CI matrix for naphthalene

5–6'	5–7'	4–6'	4–7'	5–8'	3–6'	4–8'	3–7'	3–8'
4·790	0	0	0·533	0	0	0	0	−0·272
	5·188	−0·904	0	0	0	0	0	0
		5·188	0	0	0	0	0	0
			6·395	0	0	0	0	−0·218
				5·904	0·285	0	0	0
					5·904	0	0	0
						6·852	0·595	0
							6·852	0
								8·229

include substantial contributions from electron-repulsion terms. Nevertheless, degeneracies that arise in the SCF description are precisely similar to those predicted on the basis of orbital energy differences alone and are comparable to those obtained [in $\Delta\epsilon(cm^{-1})$] by the Hückel method in Table 4-2.

Many off-diagonal elements of the matrix **h** are zero for the parent hydrocarbon, due to geometrical symmetry. The matrix can, in these circumstances, be readily factorized, and the non-zero off-diagonal elements indicate which groups of single replacement functions combine when CI methods are introduced, namely

(i) (5–6'), (4–7'), (3–8')
(ii) (5–7'), (4–6')
(iii) (5–8'), (3–6')
(iv) (4–8'), (3–7')

The last three groups illustrate first-order configuration interaction in Moffitt's[7] terminology, since they refer to degenerate configurations, and the first group demonstrates second-order CI, since the configuration energies are different. It is generally assumed that first-order effects exceed, in magnitude, those of the second order.

Diagonalization of the matrix of Table 7-2 gives the results indicated in Table 7-3, where E_j are the state energies in electron volts measured relative to the SCF ground state, and Θ_j the corresponding state wavefunctions. The matrix $\mathbf{h}$ can, for naphthalene, be easily factorized according to the groupings (i) to (iv), and the eigenvalues E_j obtained as the roots of three quadratics and one cubic.

Table 7-3

CI solution for naphthalene

j		1	2	3	4	5	6	7	8	9
r	E_j	4·284	4·618	5·619	6·092	6·188	6·257	6·508	7·446	8·287
1	5–6′	0	0·958	0	0	0	0	0·269	0	−0·098
2	5–7′	0·707	0	0	−0·707	0	0	0	0	0
3	4–6′	0·707	0	0	0·707	0	0	0	0	0
4	4–7′	0	−0·280	0	0	0	0	0·949	0	−0·141
5	5–8′	0	0	0·707	0	0·707	0	0	0	0
6	3–6′	0	0	−0·707	0	0·707	0	0	0	0
7	4–8′	0	0	0	0	0	0·707	0	0·707	0
8	3–7′	0	0	0	0	0	−0·707	0	0·707	0
9	3–8′	0	0·055	0	0	0	0	0·163	0	+0·985

Column vectors C_{rj} where $\Theta_j = \Sigma C_{rj} \Psi_r (i \to k')$

It will be observed that the lowest singlet state E_1, Θ_1 stems from interaction between the two degenerate SCF configurations $\Psi_2(5 \to 7')$ and $\Psi_3(4 \to 6')$ of energy 5·188 eV in Table 7-2 and that its partner is E_4, Θ_4. Since off-diagonal elements between this degenerate pair and all other configurations are zero, it is possible in this simple situation, to determine the CI solution from the determinant

$$\begin{vmatrix} E - 5\cdot1884 & -0.9041 \\ -0.9041 & E - 5\cdot1884 \end{vmatrix} = 0$$

giving E_1 and E_4. The corresponding state wavefunctions are the sum and difference combinations

$$\Theta_{1,4} = \frac{1}{\sqrt{2}} [\Psi_2(5 \to 7') \pm \Psi_3(4 \to 6')]$$

A similar treatment may be applied to the remaining degenerate configurations appearing on the diagonal in Table 7-2 since, with LVLS = 3 giving nine single replacement configurations, degenerate pairs are non-interacting. With LVLS = 4 giving 16 functions, and with LVLS = 5 giving the maximum number 25 of single replacement functions, degenerate pairs always interact, and the solution is more complicated than that discussed above, though the CI matrix can still be factorized to some extent because of symmetry.

It is now possible to explain the origin of the α, p, β and β' spectroscopic bands of naphthalene in the approximation (7-1) based upon the use of the nine single replacement configurations represented in Table 7-3. These bands are associated with the transition energies E_1, E_2, E_4 and E_7, in that order, to states described by the wavefunctions Θ_j ($j = 1, 2, 4$ and 7), which arise from interactions involving single replacements amongst the four innermost levels. The validity of this assignment can be confirmed by computing the associated transition intensities and polarizations for comparison with experiment (Section 7-2). The α and β bands are associated with excitations to the states Θ_j ($j = 1, 4$) in which each configuration contributes with equal weight, $|C_{rj}| = 1/\sqrt{2}$. The p and β' bands are similarly associated with excitations to Θ_2 and Θ_7 which are described mainly by the single replacement configurations $\Psi_1(5 \rightarrow 6')$ and $\Psi_4(4 \rightarrow 7')$ respectively, according to the amplitudes C_{rj} appearing in Table 7-3; both are influenced by second-order CI involving $\Psi_9(3 \rightarrow 8')$. Comparisons of the state energies E_j given in Table 7-3 and the configuration energies appearing as the diagonal elements of the matrix in Table 7-2, illustrate the characteristic 'repulsions' in computed energies when off-diagonal terms representing CIs are taken into account. These properties parallel, within the context of configuration and state energies, similar properties of repulsions amongst energy levels in Hückel and SCF theories as discussed in earlier chapters.

When LVLS = 5 the α and β bands are modified by second-order CI involving the single replacement Ψ's $(1 \rightarrow 6')$, $(5 \rightarrow 10')$, $(2 \rightarrow 8')$ and $(3 \rightarrow 9')$, and the p and β' bands are similarly modified by Ψ's $(2 \rightarrow 9')$, $(1 \rightarrow 10')$, $(1 \rightarrow 7')$ and $(4 \rightarrow 10')$.

B. Heteromolecules

It has often been assumed that the presence of heteroatoms which reduces the geometrical symmetry of a parent conjugated hydrocarbon, and as a result removes degeneracies in configuration energies, will correspondingly reduce CI to small second-order effects. However, McWeeny and Peacock[6] established that, for nitrogen derivatives of benzene, CI is still appreciable, and CI methods are essential for a valid description of excited states. A similar conclusion is reached from a consideration of N substitution in naphthalene, as illustrated in the tables given below, which may be compared with the tables already given for the parent hydrocarbon.

The results refer to N substitution in the $u = 1$ position of naphthalene (see Table 7-5) where the modification

$$\delta\omega_1 = -1\cdot66 \text{ eV} = 0\cdot70\beta \qquad [\beta = -2\cdot37 \text{ eV}]$$

proposed by McWeeny and Peacock is applied. Table 7-4 gives the SCF orbital energies for comparison with those quoted in Table 7-1.

Table 7-4

j	1	2	3	4	5
ϵ_j	−9·62	−7·74	−6·82	−5·40	−4·64

j	10	9	8	7	6
ϵ_j	9·35	7·45	6·37	5·40	4·06

The CI matrix constructed from the same single replacements given in Table 7-2, defined by taking LVLS = 3, is given in Table 7-5.

Table 7-5

CI matrix for quinolene

5–6′	5–7′	4–6′	4–7′	5–8′	3–6′	4–8′	3–7′	3–8′
4·715	0·026	0·021	0·521	0·116	0·110	0·066	−0·062	−0·244
	5·400	−0·878	0·006	0·000	−0·023	−0·165	0·010	0·149
		4·956	0·008	0·011	−0·030	0·015	−0·168	0·163
			6·401	0·079	0·073	−0·015	0·017	−0·192
				5·988	0·313	−0·007	0·050	−0·018
					5·890	0·079	−0·039	0·006
						6·677	0·577	0·057
							7·101	0·000
								8·2737

(Structure diagram of quinolene with positions labelled: 9, 10, 1, N, 2, 8, 3, 7, 5, 4, 6)

Diagonalization of the CI matrix gives the results of Table 7-6 for comparison with those of Table 7-3.

The results of Table 7-6 confirm that interaction amongst the single replacement configurations still determines, in the heteromolecule, the

Table 7-6

CI solution for quinolene

j		1	2	3	4	5	6	7	8	9
r	E_j	4·246	4·549	5·609	5·984	6·195	6·393	6·580	7·512	8·332
1	5–6′	−0·129	0·951	−0·008	0·032	−0·045	0·006	0·262	0·001	−0·084
2	5–7′	0·608	0·072	−0·069	0·677	0·086	−0·390	−0·014	−0·024	0·036
3	4–6′	0·779	0·112	0·020	−0·532	−0·083	0·295	0·033	−0·044	0·036
4	4–7′	0·020	−0·261	−0·005	−0·008	−0·401	−0·161	0·854	0·012	−0·122
5	5–8′	−0·004	−0·054	−0·635	−0·128	0·695	0·011	0·305	0·028	−0·016
6	3–6′	0·030	−0·044	0·754	0·063	0·571	0·131	0·282	0·015	−0·007
7	4–8′	0·032	−0·037	−0·119	0·397	−0·102	0·696	0·071	0·569	0·032
8	3–7′	0·035	0·041	0·089	−0·281	0·039	−0·481	−0·072	0·819	0·012
9	3–8′	−0·061	0·041	−0·002	−0·014	−0·035	−0·032	0·132	−0·024	0·987

Column vectors C_{rj} where $\Theta_j = \Sigma C_{rj}\Psi_r(i \to k')$

theoretical description of spectroscopic states. The terms which are italicised in Table 7-6, and correspond to non-zero coefficients C_{rj} in Table 7-3, remain large, and, therefore, dominate the solution as described by the excited-state wavefunctions Θ_j. Thus $C_{21} = 0.608$ and $C_{31} = 0.779$ remain close to the equal weights 0.707 ($\equiv 1/\sqrt{2}$) in the corresponding solution for the parent hydrocarbon, and the same is true of the corresponding pair $C_{24} = 0.677$ and $C_{34} = -0.532$. Reference to the CI matrix for quinolene given in Table 7-5 shows that the corresponding SCF configuration energies which lie 5.400 and 4.956 eV above the ground state separate, under CI, to give $E_1 = 4.246$ and $E_4 = 5.984$, where the largest off-diagonal element $h_{23} = -0.878$ accounts for most of the interaction. However, the two states Θ_1 and Θ_4 no longer derive exclusively from $\Psi(5 \to 7')$ and $\Psi(4 \to 6')$ but contain components from all other configurations.

The weights of components C_{rj} appearing in state wavefunctions Θ_j can generally be related to two main factors, the magnitudes of off-diagonal terms in the CI matrix which couple configurations, and the separation of configuration energies. For example, the $(4 - 8')$, $(3 - 7')$ combination ($E_6 = 6.257$ eV in Table 7-3) mixes more effectively with the neighbouring $(5 - 7')$, $(4 - 6')$ combination [$E_4 = 6.092$ eV] than with the more remote combination [$E_1 = 4.284$ eV] as indicated by the magnitudes of the relevant coefficients C_{r6} ($r = 2, 3$), C_{r1} ($r = 7, 8$) and C_{r4} ($r = 7, 8$) in Table 7-6. Indeed E_4 and E_6 suffer mutual repulsions due to this interaction so that E_6 is lowered, and E_8 raised, compared with the corresponding values in naphthalene. By tracing details of new interactions introduced by perturbations, it is usually possible to understand why some states are raised and others lowered through mutual repulsions, and why some displacements are large and others small. The coefficients italicised in Table 7-6 assist in tracing correspondences with coefficients of the solutions Θ_j for naphthalene given in Table 7-3, though the situation is not always as clear as that presented here, but can become ambiguous when mixing is heavy.

7.2 TRANSITION INTENSITIES

Some preliminary notions about the nature of the CI problem were revealed in the calculations by the Hückel method of oscillator strengths for π-electron transitions in naphthalene in Chapter 4. For example, degenerate configurations were identified, and associated intensities and polarizations were described. The resolved forms which are now quoted in Table 7-3 take sum and difference combinations, and the influence of CI on state energies can now be extended to describe corresponding effects on transition moments.

The computation of oscillator strengths in the CI method can be treated as an extension of the method described earlier for single excited configurations. Since an excited state is described as a linear combination of single replacement configurations, component transition moments, calculated between the ground state Ψ_0 and configurations $\Psi_r(i \rightarrow k')$ can be summed according to the weights C_{rj} of the linear combinations

$$\Theta_j = \sum C_{rj} \Psi_r(i \rightarrow k') \qquad (7\text{-}12)$$

to give component moments

$$Q_j^x = \int \Psi_0 x \left(\sum_r C_{rj} \Psi_r(i \rightarrow k') \right) d\tau \qquad (7\text{-}13)$$

which, by equation (4-22) transforms to

$$Q_j^x = \sqrt{2} \sum_r C_{rj} m_{ik'}^x(r) \qquad (7\text{-}14)$$

with a similar expression for Q_j^y. Oscillator strengths f_j for transitions from the ground state Ψ_0 to excited singlet states Θ_j are then calculated from the formula (4-16)

$$f_j = \left(\frac{8\pi^2 mc}{3h} \right)_{\bar{v}_j} Q_j^2$$

where (4-17)

$$Q_j^2 = (Q_j^x)^2 + (Q_j^y)^2$$

and where $_{\bar{v}_j}$ are transition energies, expressed in cm^{-1}, as computed by the CI method.

The numerical results obtained for naphthalene and quinolene which derive from the CI solutions given in Tables 7-3 and 7-6, are presented in Tables 7-7 and 7-8 respectively.

Table 7-7

Singlet excited states in naphthalene (nine configurations)

j	E_j	Q_j^x	Q_j^y	f_j	Band
1	4·284	0	0	0	α
2	4·618	0	0·553	0·247	p
3	5·619	0	0	0	
4	6·093	1·412	0	2·127	β
5	6·188	0	0	0	
6	6·257	0	0	0	
7	6·508	0	0·781	0·695	β'
8	7·446	0	0	0	
9	8·287	0	−0·804	0·938	

These results are largely self-explanatory, though they should be considered in relation to the forms of state wavefunctions Θ_j described in Table 7-3. For example, component intensities of the configurations $\Psi_2(5 \rightarrow 7')$ and $\Psi_2(4 \rightarrow 6')$ cancel in combination in the α band, and consolidate in the β band, as suggested earlier by the Hückel method (Chapter 4, Section 4). In fact the influence of CI effects upon intensities can be investigated in terms of the weights of contributing configurations, in much the same way as corresponding repulsions amongst configuration energies.

Table 7-8

Singlet excited states in quinolene (nine configurations)

j	E_j	Q_j^x	Q_j^y	f_j	Band
1	4·246	−0·222	−0·047	0·038	α
2	4·549	−0·049	0·534	0·229	p
3	5·609	−0·077	0·013	0·006	
4	5·984	1·128	0·011	1·334	β
5	6·195	0·226	−0·088	0·064	
6	6·393	−0·774	−0·060	0·675	
7	6·580	−0·029	0·822	0·779	β'
8	7·512	0·044	0·040	0·005	
9	8·332	0·012	−0·758	0·839	

The results obtained for quinolene confirm the need for CI methods of calculation even in non-degenerate situations. For example, cancellation of component intensities prevails in the α band, since the resultant intensity is small; and consolidation simultaneously persists in the β band, though with an interesting accompanying modification. The oscillator strength of the state Θ_6 of energy $E_6 = 6\cdot393$ eV in quinolene which is zero in naphthalene, takes the large value $f_6 = 0\cdot675$. The polarization of this new band is virtually along the x axis, as indicated by the magnitudes of the components Q_6^x, Q_6^y and essentially parallel to the β band. Furthermore, comparison of the state wavefunctions given in Table 7-6 indicates that in the CI description of quinolene the excited-state wavefunctions Θ_4 and Θ_6 share the same main combinations of configurations. This property implies that the oscillator strength f_6 is derived or 'borrowed' from the β band, and f_4, the oscillator strength of this band, falls accordingly.

Thus the amplitudes C_{rj} of the state wavefunctions Θ_j provide a complete and coherent account of the magnitudes of interactions amongst configurations, of the energy separations between excited states, and of polarizations and intensities of the transitions from the ground state.

7.3 APPLICATIONS OF SCF–CI METHODS

Pariser[10] has discussed theoretical properties and presented numerical results of CI methods applied to polyacenes and other conjugated molecules based upon the use of Hückel MOs. The theoretical formulation of the CI problem is similar to that described earlier in this chapter, though the matrix elements of $\mathbf{h}$ are rather more complicated than those based upon the use of SCF orbitals. SCF–CI methods are, in fact, becoming increasingly applied through the use of digital computers, but it is premature at this point in time, and undesirable in the present context, to make an assessment of the method. Instead, a few typical applications will be introduced in terms of the input data required for computing solutions by the programs presented at the end of the chapter, and some brief comments will be attached in notes that follow. Many of the problems that have been discussed in previous chapters, and most applications to conjugated molecules which are described, for example, in Streitwieser's book, can be reconsidered in terms of SCF–CI theory.

The SCF–CI computer programs are easy to use, through the simple form of input data, which is based upon the use of the hexagonal grid (Figure 4-21) of Chapter 4, and automatic selection of configurations by the parameter setting LVLS. Practical experience in the study of solutions obtained from the computer can provide a basis for understanding features and properties that characterize the π-electron MO CI method, as indicated in previous sections. It is useful, therefore, to begin to experiment with the computer programs by generating solutions for benzene, naphthalene, anthracene, phenanthrene[11] and other 'prototype' AHs and related nitrogen heterocyclics; these examples will provide a useful foundation for further studies.

Data Set

The following data set computes SCF–CI solutions for

   (a) naphthalene, quinolene, and isoquinolene

   (b) aniline

   (c) cyclooctadecanonaene $C_{18}H_{18}$

X1 card	
X2 „	
Y1 „	grid coordinates
Y2 „	

(a) 0031	NMOLS, NSPEC	
	(NMOLS = number of parent hydrocarbons)	
010005	N, M	
0102030405060110111213	NATM	'naphthalene' grid
0005	LVLS	
0003	NDER	
0	LAB	parent hydrocarbon
1	LAB	quinolene; modify F
01	NITEM	number of modified elements
0101 − 01·659	I, J, F(I, J)	$\delta\omega_1 = 0\cdot7\beta$ ($\beta = -2\cdot37$)
0	LAB	
1	LAB	isoquinolene
01	NITEM	
0202 − 1·659	I, J, F(I, J)	$\delta\omega_2 = 0\cdot7\beta$
0	LAB	

(b)	007004	N, M	
	01020304050607	NATM	'benzyl' grid
	0003	LVLS	
	0002	NDER	
	1	LAB	modify F
	01	NITEM·	number of modified elements
	0707 − 15·000	I, J, F(I, J)	$\delta\omega_7 = -15$ eV
	2	LAB	modify G
	01	NITEM	number of modified elements
	0707 + 14·090	I, J, G(I, J)	$\gamma_{77} = 14\cdot09$ eV
	3	LAB	modify Z
	01	NITEM	number of modified elements
	072·0	J, Z(J)	$Z_7 = 2$
	0	LAB	no further modifications
	1		second aniline calculation
	02		with the same parameters
	0707 − 15·000		except:
	0701 − 01·896		$\beta_{71} = 0\cdot8\beta$
	2		
	01		
	0707 + 14·090		
	3		
	01		
	072·0		
	0		
(c)	018009	N, M	
	070809101112131415161718192021222324		'C$_{18}$H$_{18}$' grid
	0004	LVLS	
	0001	NDER	
	0	LAB	parent hydrocarbon.

Notes on the calculations

(a) The setting LVLS = 5 calculates excited states by permitting interaction amongst all 25 (=MINK) single replacement configurations. This produces a large output which may be useful for analytical purposes, but could be restricted, by program modification, to select for printing only states of physical significance.

Note that in each of the 3 (=NDER) calculations, for naphthalene, quinolene, and isoquinolene, the set of modification cards is terminated by LAB = 0. The modification $\delta\omega_N$ in core parameter, given by $\delta\omega_N = 0\cdot7\beta = 1\cdot659$ eV corresponds to the value proposed by McWeeny and Peacock.[6]

(b) Two calculations are performed for aniline with the sets of parameters

(i) $\delta\omega_N = -15$ eV
        $\gamma_{NN} = 14\cdot09$ eV
        $Z_N = 2$

(ii) $\delta\omega_N = -15$ eV
        $\beta_{N-C} = 0\cdot8\beta = -1\cdot896$ eV
        $\gamma_{NN} = 14\cdot09$ eV
        $Z_N = 2$

Since the core integral ω for a conjugated carbon atom is about -11 eV the value of ω_N is equivalent to about -26 eV, roughly in agreement with the values proposed by Dewar and Paolini,[8] and by Bloor et al.[9] The electron-repulsion integral $\gamma_{NN} = 14\cdot09$ eV corresponds to that suggested by Dewar.

The first calculation for aniline assumes $\beta_{N-C} = \beta$ and the second introduces the modification $\beta_{N-C} = 0\cdot8\beta$, and is assumed to represent a reduced overlap between the nitrogen $\phi(2p_z)$ atomic orbital, and that of the adjacent carbon atom, due to non-planarity.

(c) Excited states are calculated from the 16 configurations defined by taking LVLS = 4. SCF levels ϵ_j occur in degenerate pairs in the cyclic polyenes $C_{4n+2}H_{4n+2}$ except for the lowest, $j = 1$, and highest $j = 4n + 2$,

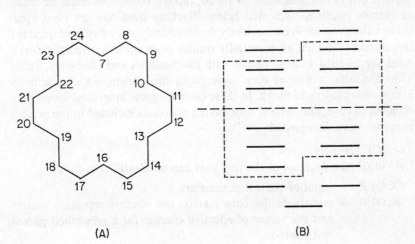

(A)   (B)

as represented diagrammatically in (b) for the case $n = 4$. Invalid results will be obtained by taking odd-integral values for LVLS except when all levels are to be taken into account (in this case LVLS = 9), since the subroutine SCOFA must then select one configuration from the outermost degenerate pairs defined by the value of LVLS. The situation is illustrated by the enclosed region in (a) for the case LVLS = 5, which leads to an arbitrary selection of one, and only one of the outermost degenerate pairs which could, in principle, be included. A valid solution can be obtained only if all configurations associated with a degenerate situation are included in the expansion form (7-1).

7.4 COMPUTER PROGRAMS (p. 197)

The MAIN program that calls the set of subroutines solving the SCF–CI equations for conjugated molecules falls into two distinct parts. The first part terminates at the FORTRAN statement

82   GO TO   76

and obtains solutions of the SCF equations; the second part, represented by the rest of the program solves the CI problem. The two parts are largely independent; the first transmits to the second eigensolutions of the SCF equations, and atom coordinates that are used in calculating both dipole and transition moments.

The program as it stands requires in all about 18,000 words of 24 bit core store; this includes the storage of data and intermediate results in various arrays of dimensions 30 by 30, though economies could be made at various points as indicated below. Backing store has not been used, though the program is conveniently designed and can be easily adapted for this purpose; in fact, an essentially similar program was originally developed for an IBM 1620 computer with disc facilities and effectively, about 1200 available words of core store, with the dimensions of the main arrays restricted to 15 by 15. In these circumstances innermost loops were retained in core, and certain time-saving processes included in the present program were disregarded.

The SCF first part
Individual subprograms of the first part can be identified as follows

    SCOFZ — input of control parameters.

    SCOFB — constructs the core matrix, the electron-repulsion matrix and the vector of effective charges for a prescribed parent hydrocarbon.

SCOFD — initiates these same matrices for processing for each new derivative.

SCOFH — modifies selected array elements for the prescribed molecule.

SCOFI1 — JACOBI diagonalization routine.

SCOFI2 — orders eigenvalues and vectors.

SCOFI3 — constructs new F or $\mathbf{h}'$ matrix.

SCOFI4 — preliminary transformation of F.

SCOFJ — general purpose print routine.

SCOFK — computes SCF π electron energy E, bond-order matrix $\mathbf{P}$, dipole moment and components.

SCOFA — though entered in the first part of the program, belongs to the CI calculation, and is described below.

SCOFZ

This routine reads, as input parameters

NMOLS — the number of different parent hydrocarbons.

NSPEC   — zero for SCF only; otherwise set equal to unity.

The following parameters are preset in the program

BETA $= -2\cdot37$; the framework resonance integral, in eV

HAFGAM $= \gamma/2$ where $\gamma = 11\cdot35$ eV is the carbon electron-repulsion integral

EPS $= 1 \times 10^{-16}$; terminating criterion for the matrix diagonalization routine.

SCOFB

This routine generates the core framework matrix, the matrix of electron-repulsion integrals, and the effective nuclear charges for a parent AH based upon the hexagonal-grid structure of Figure 4-21. It is then possible to change elements of any arrays, in the modification routine SCOFH, to describe a molecule under investigation.

SCOFB greatly simplifies the input specification by generating the bulk of input data internally. The procedure is essentially similar to that described in Chapter 4. Hexagonal-grid coordinates are read and held permanently in the arrays IX, IY in the main program. The vector NATM which enumerates a set of grid atoms defining the parent hydrocarbon is read in SCOFB, preceded by a specification N of the number of atoms and M, the number of orbitals occupied in the ground state. Grid coordinates of the selected set are converted to molecular coordinates XCRD, YCRD, and D, the distance between all pairs of atoms is computed. Matrix elements $G(I, J)$ are then computed as follows

    (i) if $D > 2.81$ $G(I, J) = 14.4/D$
    (ii) if $2.81 > D > 2.75$ $G(I, J) = 4.79$
    (iii) if $2.75 > D > 1.42$ $G(I, J) = 5.77$
    (iv) if $1.42 > D$ $G(I, J) = 7.19$

and, when $I = J$, $G(I, J) = 11.35$. The bounds used in (i) to (iv) ensure that the values of γ_{rs} expressed in electron volts, as proposed by Parr and Pariser for hexagonally disposed atoms r and s within the same or neighbouring rings are introduced correctly.

Values of γ_{rs} for the parent hydrocarbon are preserved in the lower semimatrix of **FG**; the upper half excluding the diagonal preserves the corresponding core integrals. All elements of the upper half are put equal to zero except those for which $G(I, J) > 7 \, eV$ which identifies nearest neighbours, in which case the value $-2.37 \, eV$ is assigned.

The molecular coordinates XCRD, YCRD are identified with the arrays X and Y in the main program.

SCOFD

The core matrix **F**, and electron-repulsion integrals matrix **G** of a parent hydrocarbon are extracted from **FG** to initiate calculations for a new 'derivative'. Zero-diagonal elements of the core matrix are assigned to the vector FDIAG, and all elements of the vector $Z(I)$ of effective nuclear charges are initially assigned the value unity.

SCOFH   Matrix elements of the arrays **F**, **G** (=**GAM**) and **Z** may be modified to 'convert' a real or hypothetical parent hydrocarbon to the molecule under investigation.

Modifications may be specified by reading data on cards according to the following prescription.

Format	Parameters	Value	Result
I1	LAB	0	Exit from routine.
		1	Modify F.
		2	„ GAM.
		3	„ Z.
I2	NITEM		Number of items to be modified for each value of LAB.
2I2, F7.3	I, J, X		For LAB = 1, 2; sets (I, J)th element to the value X.
I2, F3.1	J, Z(J)		For LAB = 3; inputs Z(J).

Each set of modifications must be termined by LAB = 0 to ensure exit from the subroutine. If the solution for the parent hydrocarbon itself is required, the input LAB = 0 should be applied as a separate entry. The modifications appear on the printed output under headings F (or G, or Z) MODIFICATIONS.

SCOFJ is a generalized print routine.

SCOFI1 is the standard JACOBI matrix diagonalization routine described earlier. The lower semimatrix of A (or F) including **ADIAG** (or **FDIAG**) only, is diagonalized. (See Chapter 2.)

SCOFI2 arranges eigenvalues in ascending order and eigenvectors correspondingly.

SCOFI3 constructs a new matrix $\mathbf{h}'$ from the previous diagonalization according to the formulae

$$h_{rr} = \omega_r + \tfrac{1}{2}(P_{rr}\gamma_{rr} - \gamma) + \sum_{t \neq r}(P_{tt} - Z_t)\gamma_{rt}$$

$$h_{rs} = \beta_{rs} - \tfrac{1}{2}P_{rs}\gamma_{rs}$$

where the diagonal elements h_{rr} are stored in **FDIAG** and h_{rs} in the lower triangular part of F. The core framework (ω_r, β_{rs}) matrix for the molecule under investigation is preserved in the upper part of **F**.

SCOFI4 is a procedure, apparently first suggested by Bloor et al.,[9] for speeding up the SCF iteration process. The JACOBI diagonalization process is normally initiated by setting **U**, the matrix of column vectors, equal to the unit matrix. Since approximations **W** to the matrix of eigenvectors are available at each iteration step, the transformation

$$\mathbf{W}\dagger\mathbf{h}'\mathbf{W} \to \mathbf{D}'$$

brings $\mathbf{D}'$ into approximately diagonal form, and the process

$$\dots \mathbf{T}_t^\dagger \dots \mathbf{D}' \dots \mathbf{T}_t \dots$$

of two-dimensional rotations $\mathbf{T}_t$ required to complete diagonalization to the prescribed accuracy is appreciably reduced.

The subroutine computes the matrix product $\mathbf{h}'\mathbf{W}$, where F at this stage represents $\mathbf{h}'$, and T represents **W**, to temporary storage in the array (FS), and subsequently restores the product $\mathbf{T}\dagger$(FS) to F; the lower part of F only, is involved in these transformations. Note that on reentry to the diagonalization routine SCOFI1, the parameter NIT is not zero and, therefore, the program segment setting **U** = **I**, the unit matrix, is bypassed, and A is identified with $\mathbf{D}'$.

SCOFK. This subroutine calculates and prints the energy E of the π electrons in the SCF approximation, the density matrix **P**, and the dipole moment with its components, following completion of the iterative solution of the SCF equations.

The total π-electron energy E is computed, in effect, from Pople's formula

$$E = \tfrac{1}{2}\sum\sum P_{rs}(F_{rs} + H_{rs}^{core})$$

The computed value is printed alongside the heading

BINDING ENERGY $E = \ldots$ value $\ldots$

and this is followed by

DIPOLE MOMENT $= \ldots$ value $\ldots$ XMU $= \ldots$ value $\ldots$ YMU $= \ldots$ value $\ldots$

where the values associated with XMU and YMU are the components μ^x and μ^y measured parallel to the coordinates of the grid framework of Figure 4-21. Next the density matrix **P** is printed in semimatrix form, under the heading

DENSITY MATRIX

P_{11}

P_{12} P_{22}

P_{13} P_{23} P_{33}

$\ldots$ $\ldots$ $\ldots$ $\ldots$

etc.

The SCF energy levels ϵ_j $(j = 1, 2, \ldots N)$ and MOs

$$\psi_j = \sum c_{rj}\phi_r \qquad (r = 1, 2, \ldots N)$$

are then printed, as follows, by the subroutine SCOFJ, under the heading

SCF LEVELS AND ORBITALS

ϵ_1

c_{11} c_{21} c_{31} c_{41} $\ldots$ $\ldots$

$\ldots$ $\ldots$ etc.

ϵ_2

c_{12} c_{22} c_{32} c_{42} $\ldots$ $\ldots$

$\ldots$ $\ldots$ etc.

and so on, where the energy levels, and associated orbitals, are printed in ascending order with ϵ_1 the lowest, most bound level.

The CI second part

The subprograms can be identified as follows:

SCOFA — selects single replacement configurations

SCOFI5 — computes components $m_{ik'}^z$ and $m_{ik'}^y$ of the transition moment between the ground state Ψ_0 and each single replacement configuration $\Psi'(i \to k')$

SCOFL — constructs CI matrices for singlet and triplet excitations

SCOFI1 — standard JACOBI diagonalization of CI matrices

SCOFI2 — arranges eigenvalues in ascending order, and eigenvectors accordingly

SCOFM — prints CI solution; computes and prints transition moments and oscillator strengths

SCOFP — prints energies of configurations in the SCF approximation

SCOFA. This subroutine is essentially similar to the first part of TRMOM introduced in Chapter 4. The value of LVLS specifies the number of highest occupied orbitals i and lowest unoccupied orbitals k' of the SCF ground state Ψ_0 from which single replacement configurations $\Psi'(i \to k')$ are to be selected; these are MINK $=$ (LVLS)^2 in number.

SCOFI5 computes the components $m_{ik'}^z$ and $m_{ik'}^y$ for each of the MINK transitions selected by SCOFA. This step must be made before entry to the CI calculation, which destroys the matrix $\mathbf{C}$ of SCF MOs from which the components are calculated.

SCOFL. The routine computes matrix elements between single replacement configurations selected by SCOFA, for both singlet and triplet forms according to the formulae (7-8) and (7-10). The CI matrix for singlet states is constructed in the lower triangular part of the array AS, and that for triplets in AT; AS is formally identified with $\mathbf{F}$ in the argument list of the calling statement.

SCOFP. Diagonal elements (equation 7-8) of the CI matrices represent the energies of configurations $\Psi'(i \to k')$ in the SCF approximation, measured relative to the ground-state configuration. These are printed in SCOFP for both singlet and triplet configurations under the two headings

SCF SINGLET (or TRIPLET) CONFIGURATIONS

Values of the diagonal elements are printed below corresponding pairs of labels $i - k'$ defining the configurations, thus

$$\ldots i - k' \ldots \text{etc.}$$
$$\ldots \text{value} \ldots \text{etc.}$$

SCOFI1. Diagonalizes the matrix **AS** through identification with **F**, operating on the lower semimatrix only. (See above and Chapter 2.)

SCOFM. The state energies E_j obtained as eigenvalues in SCOFI1, and ordered in SCOFI2, are printed, with corresponding state wavefunctions

$$\Theta_j = \sum C_{tj}\Psi_j(i \to k') \tag{7-12}$$

obtained as eigenvectors, under the two headings

SINGLET (or TRIPLET) STATES

Each state energy E_j is printed in the form

ENERGY = value

and is followed by the corresponding state wavefunction Θ_j, where the values of the coefficients C_{tj} are tabulated under the corresponding labels i, k' identifying the configurations, thus

$\ldots i - k' \ldots$ etc.
$\ldots$ value $\ldots$ etc.

Transition moments are computed in terms of the linear combinations given in (7-14), where the components $m_{ik'}^z$, $m_{ik'}^y$ are computed earlier in SCOFI5, and the weights C_{tj} have now been derived. Oscillator strengths are then computed according to the formula (4-16). The results are tabulated under the heading

OSCILLATOR STRENGTHS

where the columns X-COMPONENT and Y-COMPONENT refer (equation 7-14) to Q^x and Q^y respectively.

Computed results for singlet states are printed in SCOFM the first time round. Then the CI matrix **AT** for triplet states is transferred to the lower semimatrix of **AS** which in turn is identified with **F** in the argument list of the calling statement. Thus the results for triplet states are printed in SCOFM the second time round. The integer variable I2 is used to control these two loops.

It should be emphasized that orthogonality of the spin functions ensures that the oscillator strengths for all excitations from the ground state to the tabulated triplet states are zero, and that the corresponding transitions are, therefore, forbidden. The values tabulated in this section of the program output refer only to the orbital component. It is, however, not wholly irrational to print these results since in certain physical situations the transitions can, in fact, be observed. For example, in the oxygen enhancement experiments of Evans[12,13] induced singlet-triplet transitions arise from spin-orbit perturbations, due to the inhomogeneous field of the paramagnetic oxygen molecule. The strongest absorptions occur when the transitions are orbitally allowed and only spin-forbidden. The calculated results are, therefore, not entirely without physical significance, but may be disregarded at the reader's discretion.

The MAIN Program

A prescribed series of π-electron calculations is controlled by the MAIN program. The control variable KMOLS labels and counts the set of NMOLS parent hydrocarbons selected from the grid framework of Figure 4-21 where the value NMOLS is read in SCOFZ. A prescribed number of 'derivatives' is to be associated with each value of KMOLS, that is, with each parent hydrocarbon. This number is read into NDER which controls the loop beginning

$$\text{DO } 83 \text{ I} = 1, \text{NDER}$$

that encloses the SCF–CI calculation.

Solutions for parent hydrocarbons must be included in the number of derivatives NDER, and associated with the parameter value LAB = 0 in SCOFH.

The SCF iteration process, controlled by the count NIT, is terminated after 10 iterations. It would be preferable to apply a genuine theoretical test for termination. Such a test would involve several matrix operations, to be applied at each iteration stage and would, therefore, consume valuable computing time and storage. In practice the final few iterations are always executed rapidly, following partial diagonalization in SCOFI4, and an increase to 13 or 15 iterations instead of 10, would consume less time than computing a theoretical criterion. Ten iterations is found acceptable in most cases, but is recognizably a weak criterion based only on practical experience.

An extrapolation procedure introduced in an earlier program reduced the computing time by, roughly, a factor of two. It has not been included here because it consumes storage space, and may be less effective when Bloor's procedure (SCOFI4) is introduced.

The calculation of state energies and wavefunctions in the CI second part is controlled by a switch I2 that is set in SCOFL and incremented in SCOFM. Singlet and triplet states are computed when I2 takes the values 0 and 1 respectively, and I2 = 2 exits to the end of the loop in the MAIN program.

A. Results

A specimen set of results follows for aniline, based on the data specification (b) given in the previous section, with the restriction that NDER $\equiv$ 0001.

```
CORE INTEGRALS
 0.000
-2.370  0.000
 0.000 -2.370  0.000
 0.000  0.000 -2.370  0.000
 0.000  0.000  0.000 -2.370  0.000
-2.370  0.000  0.000  0.000 -2.370  0.000
-2.370  0.000  0.000  0.000  0.000  0.000  0.000

ELECTRON-REPULSION INTEGRALS
11.350
 7.190 11.350
 5.770  7.190 11.350
 4.970  5.770  7.190 11.350
 5.770  4.970  5.770  7.190 11.350
 7.190  5.770  4.970  5.770  7.190 11.350
 7.190  5.770  3.888  3.429  3.888  5.770 11.350

F MODIFICATIONS
7 7-15.000
G MODIFICATIONS
7 7 14.090
Z MODIFICATIONS
7  2.0

BINDING ENERGY E= -0.18159138E+03

DIPOLE MOMENT=  1.2573   XMU= -0.0000   YMU= -1.2573

DENSITY MATRIX
 0.96754
 0.63105   1.06331
 0.00627   0.67146   0.98987
-0.31470  -0.04007   0.66373   1.02716
 0.00627  -0.32831   0.01108   0.66373   0.98987
 0.63105   0.04209  -0.32831  -0.04007   0.67146   1.06331
 0.32155  -0.18906  -0.02324   0.12621  -0.02324  -0.18906   1.89895

SCF LEVELS AND ORBITALS

-0.9377E+01

 0.535827   0.330287   0.226717   0.186358   0.226717   0.330287   0.597645

-0.7510E+01

 0.080371  -0.202928  -0.399686  -0.476303  -0.399686  -0.202928   0.604001

-0.5419E+01

 0.000000  -0.505276  -0.494667  -0.000000   0.494667   0.505276  -0.000000

-0.4564E+01

-0.436119  -0.355078   0.197706   0.501981   0.197706  -0.355078   0.476947

 0.5783E+01

 0.000000  -0.494667   0.505276   0.000000  -0.505276   0.494667  -0.000000

 0.5997E+01

-0.581409   0.252253   0.297140  -0.566251   0.297140   0.252253   0.193585

 0.8975E+01

-0.422130   0.400023  -0.401833   0.407161  -0.401833   0.400023   0.114233

C.I.SOLUTION


SCF SINGLET CONFIGURATION ENERGIES

 4 - 5     4 - 6     3 - 5     3 - 6     4 - 7     2 - 5     3 - 7  ·  2 - 6     2 - 7
 5.1959    5.8535    6.3615    5.9203    7.9017    7.8509    8.4749    8.1369   10.9449
```

SINGLET STATES

ENERGY = 0.451950E+01

4 - 5	4 - 6	3 - 5	3 - 6	4 - 7	2 - 5	3 - 7	2 - 6	2 - 7
0.8161	0.0000	0.0000	0.5674	0.0000	-0.1098	-0.0044	0.0000	0.0000

ENERGY = 0.554854E+01

4 - 5	4 - 6	3 - 5	3 - 6	4 - 7	2 - 5	3 - 7	2 - 6	2 - 7
0.0000	0.8690	-0.4814	0.0000	0.0829	0.0000	0.0000	0.0252	0.0753

ENERGY = 0.654037E+01

4 - 5	4 - 6	3 - 5	3 - 6	4 - 7	2 - 5	3 - 7	2 - 6	2 - 7
-0.5771	0.0000	0.0000	0.8097	0.0000	-0.1059	0.0125	0.0000	0.0000

ENERGY = 0.660137E+01

4 - 5	4 - 6	3 - 5	3 - 6	4 - 7	2 - 5	3 - 7	2 - 6	2 - 7
0.0000	0.4722	0.8746	0.0000	0.0872	0.0000	0.0000	0.0613	0.0259

ENERGY = 0.763403E+01

4 - 5	4 - 6	3 - 5	3 - 6	4 - 7	2 - 5	3 - 7	2 - 6	2 - 7
0.0000	-0.1212	-0.0546	0.0000	0.7939	0.0000	0.0000	0.5929	-0.0232

ENERGY = 0.787757E+01

4 - 5	4 - 6	3 - 5	3 - 6	4 - 7	2 - 5	3 - 7	2 - 6	2 - 7
0.0258	0.0000	0.0000	0.1479	0.0000	0.9648	-0.2161	0.0000	0.0000

ENERGY = 0.835547E+01

4 - 5	4 - 6	3 - 5	3 - 6	4 - 7	2 - 5	3 - 7	2 - 6	2 - 7
0.0000	0.0130	-0.0092	0.0000	-0.5855	0.0000	0.0000	0.7925	0.1699

ENERGY = 0.850453E+01

4 - 5	4 - 6	3 - 5	3 - 6	4 - 7	2 - 5	3 - 7	2 - 6	2 - 7
0.0168	0.0000	0.0000	0.0249	0.0000	0.2144	0.9763	0.0000	0.0000

ENERGY = 0.110590E+02

4 - 5	4 - 6	3 - 5	3 - 6	4 - 7	2 - 5	3 - 7	2 - 6	2 - 7
0.0000	-0.0842	0.0141	0.0000	0.1114	0.0000	0.0000	-0.1266	0.9820

OSCILLATOR STRENGTHS

ENERGY	X-COMPONENT	Y-COMPONENT	OSCILLATOR STRENGTH
0.451950E+01	0.195	-0.000	0.030
0.554854E+01	0.000	0.370	0.133
0.654037E+01	-0.898	8.000	0.924
0.660137E+01	0.000	0.996	1.146
0.763403E+01	-0.000	-0.121	0.020
0.787757E+01	-0.317	0.000	0.139
0.835547E+01	-0.000	-0.021	0.001
0.850453E+01	-0.066	0.000	0.007
0.110590E+02	0.000	0.026	0.001

SCF TRIPLET CONFIGURATION ENERGIES

4 - 5	4 - 6	3 - 5	3 - 6	4 - 7	2 - 5	3 - 7	2 - 6	2 - 7
3.9730	3.7824	3.8826	4.7427	6.9448	7.0757	7.3947	7.0834	10.1734

TRIPLET STATES

ENERGY = 0.319992E+01

4 - 5	4 - 6	3 - 5	3 - 6	4 - 7	2 - 5	3 - 7	2 - 6	2 - 7
0.0000	0.7708	-0.5900	0.0000	0.0495	0.0000	0.0000	-0.1262	0.1987

ENERGY = 0.381790E+01

4 - 5	4 - 6	3 - 5	3 - 6	4 - 7	2 - 5	3 - 7	2 - 6	2 - 7
0.9505	0.0000	0.0000	-0.2697	0.0000	-0.1527	-0.0227	0.0000	0.0000

ENERGY = 0.405798E+01

4 - 5	4 - 6	3 - 5	3 - 6	4 - 7	2 - 5	3 - 7	2 - 6	2 - 7
0.0000	0.6000	0.7955	0.0000	-0.0218	0.0000	0.0000	-0.0814	-0.0117

ENERGY = 0.480519E+01

4 - 5	4 - 6	3 - 5	3 - 6	4 - 7	2 - 5	3 - 7	2 - 6	2 - 7
0.2748	0.0000	0.0000	0.9598	0.0000	0.0232	-0.0529	0.0000	0.0000

ENERGY = 0.588929E+01

4 - 5	4 - 6	3 - 5	3 - 6	4 - 7	2 - 5	3 - 7	2 - 6	2 - 7
0.0000	0.0931	0.0157	0.0000	0.7480	0.0000	0.0000	0.6510	-0.0875

ENERGY = 0.631079E+01

4 - 5	4 - 6	3 - 5	3 - 6	4 - 7	2 - 5	3 - 7	2 - 6	2 - 7
0.0815	0.0000	0.0000	-0.0777	0.0000	0.7428	-0.6600	0.0000	0.0000

ENERGY = 0.812830E+01

4 - 5	4 - 6	3 - 5	3 - 6	4 - 7	2 - 5	3 - 7	2 - 6	2 - 7
0.0000	0.1015	-0.0153	0.0000	-0.6353	0.0000	0.0000	0.7414	0.1902

ENERGY = 0.825220E+01

4 - 5	4 - 6	3 - 5	3 - 6	4 - 7	2 - 5	3 - 7	2 - 6	2 - 7
0.1201	0.0000	0.0000	-0.0088	0.0000	0.6514	0.7491	0.0000	0.0000

ENERGY = 0.105910E+02

4 - 5	4 - 6	3 - 5	3 - 6	4 - 7	2 - 5	3 - 7	2 - 6	2 - 7
0.0000	-0.1643	0.1366	0.0000	0.1840	0.0000	0.0000	-0.0626	0.9574

OSCILLATOR STRENGTHS

ENERGY	X-COMPONENT	Y-COMPONENT	OSCILLATOR STRENGTH
0.319992E+01	0.000	0.227	0.029
0.381790E+01	0.852	-0.000	0.486
0.405798E+01	0.000	1.039	0.767
0.480519E+01	-0.460	0.000	0.178
0.588929E+01	-0.000	0.093	0.009
0.631079E+01	-0.071	0.000	0.006
0.812830E+01	-0.000	0.047	0.003
0.825220E+01	-0.080	0.000	0.009
0.105910E+02	0.000	0.047	0.004

END

&END;
TIME = 0003 31
    A

B. Listings

```
      MAINO 2
      DIMENSION F(30,30),FDIAG(30),T(30,30),G(30,30),Z(30),AT(30,30)
      DIMENSION IX(64),IY(64)
      DIMENSION LL(64),LH(64),FG(30,30),X(96),Y(96)
      DIMENSION XMIK(64),YMIK(64)
      READ(7,98)(IX(I),I=1,54)
      READ(7,98)(IY(I),I=1,54)
   98 FORMAT(40I2)
      KMOLS=0
      CALL SCOFZ(NMOLS,NSPEC,EPS,HAFGAM,BETA)
   76 KMOLS=KMOLS+1
      IF(KMOLS-NMOLS)77,77,78
   77 CONTINUE
      CALL SCOFB(N,M,FG,G,Z,X,Y,IX,IY)
      CALL SCOFA(MINK,LL,LH,N)
      READ(7,99)NDER
   99 FORMAT(I4)
      DO 83 I=1,NDER
      CALL SCOFD(N,F,FG,FDIAG,G,Z)
      MIND=6
      CALL SCOFJ(N,MIND,FG,FDIAG,T,G,Z)
      CALL SCOFH(N,F,FDIAG,G,Z)
      NIT=0
   17 CALL SCOFI1(N,F,FDIAG,T,NIT,EPS)
      CALL SCOFI2(N,F,FDIAG,T)
      CALL SCOFI3(N,M,F,FDIAG,T,G,Z,HAFGAM)
      CALL SCOFI4(N,F,FDIAG,T)
      NIT=NIT+1
      IF (NIT-10)17,17,18
   18 CALL SCOFK(N,M,F,T,G,Z,X,Y)
      CALL SCOFI5(MINK,LL,LH,N,XMIK,YMIK,T,X,Y)
      MIND=3
      CALL SCOFJ(N,MIND,F,FDIAG,T,G,Z)
      IF(NSPEC)81,82,81
   82 GO TO 76
   81 CALL SCOFL (N,MINK,F,FDIAG,T,G,AT,I2,LL,LH)
      WRITE(2,1000)
 1000 FORMAT(/12HC.I.SOLUTION)
   84 NIT =0
      CALL SCOFP(MINK,LL,LH,FDIAG,I2)
      CALL SCOFI1(MINK,F,FDIAG,T,NIT,EPS)
      CALL SCOFI2(MINK,F,FDIAG,T)
      CALL SCOFM(MINK,F,AT,FDIAG,T,I2,XMIK,YMIK,LL,LH)
      IF (I2-2)84,83,83
   83 CONTINUE
      GO TO 76
   78 STOP
      END




      SUBROUTINE SCOFZ(NMOLS,NSPEC,EPS,HAFGAM,BETA)
      EPS=1.0E-15
      HAFGAM=5.675
      BETA=-2.37
      READ(7,100)NMOLS,NSPEC
  100 FORMAT(I3,I1)
      RETURN
      END
```

```
      SUBROUTINE SCOFB(N,M,FG,G,Z,XCRD,YCRD,IX,IY)
      DIMENSION FG(30,30),G(30,30),Z(30)
      DIMENSION IX(64),IY(64),NATM(96)
      DIMENSION XCRD(96),YCRD(96)
      DIMENSION X(96),Y(96)
      EL=1.40
      XT=0.8660254*EL
      YT=0.5*EL
      DO 4 I=1,54
      X(I)=IX(I)*XT
    4 Y(I)=IY(I)*YT
      READ(7,99)N,M
      READ(7,98) (NATM(J),J=1,N)
      DO 9 I=1,N
      DO 9 J=1,N
    9 FG(I,J)=0.
      DO 10 I=1,N
      ID=NATM(I)
      XD = X(ID)
      YD = Y(ID)
      YCRD(I)=YD
      XCRD(I)=XD
      DO 10 J=I,N
      IF(I-J)12,13,12
   13 G(I,I)=11.35
      GO TO 10
   12 JD=NATM(J)
      XC=X(JD)
      YC=Y(JD)
      D=SQRT((XC-XD)**2+(YC-YD)**2)
      IF(D-2.81)14,14,51
   51 G(I,J)=14.4/D
      GO TO 10
   14 IF(D-2.75)15,15,52
   52 G(I,J)=4.97
      GO TO 10
   15 IF(D-1.42)16,16,17
   17 G(I,J)=5.77
      GO TO 10
   16 G(I,J)=7.19
   10 CONTINUE
      DO 19 I=1,N
      DO 19 J=I,N
      FG(I,J)=G(I,J)
   19 G(J,I)=G(I,J)
      IUPP=N-1
      DO 20 I=1,IUPP
      JLOW=I+1
      DO 20 J=JLOW,N
      IF(G(I,J)-7.0)20,20,21
   21 FG(J,I)=-2.37
   20 CONTINUE
      DO 113 J=1,N
  113 Z(J)=1.0
   98 FORMAT(40I2)
   99 FORMAT(2I3)
      RETURN
      END
```

```
      SUBROUTINE SCOFA(MINK,LL,LH,N)
      DIMENSION LL(64),LH(64)
      READ(7,99)LVLS
      IF(LVLS.EQ.0)GO TO 13
 99   FORMAT(I4)
      M=(N+1)/2
      I=1.
      DO 10 J=1,LVLS
      MD=M+1
      MT=M
      IF(J-1)17,17,11
 17   LH(I)=MD
      LL(I)=MT
      GO TO 10
 11   KUPP=J-1
      DO 12 K=1,KUPP
      I=I+1
      LH(I)=M+J
      LL(I)=MT
      I=I+1
      LH(I)=MD
      LL(I)=M-KUPP
      MT=MT-1
 12   MD=MD+1
      I=I+1
      LL(I)=M-KUPP
      LH(I)=M+J
 10   CONTINUE
      MINK=LVLS*LVLS
 13   CONTINUE
      RETURN
      END




      SUBROUTINE SCOFD(N,F,FG  ,FDIAG,GAM,Z)
      DIMENSION F(30,30),FG  (30,30),FDIAG(30),GAM(30,30),Z(30)
      DO 10 J=1,N
      DO 10 I=J,N
      IF(I-J)16,17,16
 16   F(I,J)=FG  (I,J)
      F(J,I)=F(I,J)
      GO TO 10
 17   FDIAG(J)=0
      F(J,J)=0
 10   CONTINUE
      DO 11 I=1,N
      DO 11 J=I,N
      GAM(I,J)=FG  (I,J)
 11   GAM(J,I)=GAM(I,J)
      DO 12 J=1,N
 12   Z(J)=1.0
      RETURN
      END
```

```
      SUBROUTINE SCOFJ(N,MIND,A,ADIAG,U,G,Z).
      DIMENSION A(30,30),ADIAG(30),U(30,30),G(30,30),Z(30)
      GO TO (9,10,11,12,13,14),MIND
   10 DO 19 J=2,N
      IUP=J-1
   19 WRITE(2,100) (A(I,J),I=1,IUP)
      WRITE(2,100) (ADIAG(J),J=1,N)
      GO TO 9
   11 WRITE (2,106)
  106 FORMAT    (/24H SCF LEVELS AND ORBITALS)
      DO 17 J=1,N
      WRITE(2,112)ADIAG(J)
  112 FORMAT (1H0,1X,E11.4/)
   17 WRITE(2,111) (U(I,J),I=1,N)
  111 FORMAT(7(F11.6))
      GO TO 9
   12 WRITE(2,101) ((U(I,J),I=1,N),J=1,N)
      WRITE(2,100)(ADIAG(I),I=1,N)
      GO TO 9
   13 WRITE(2,103) ((A(I,J),I=1,J),J=1,N)
      WRITE(2,103) ((G(I,J),I=1,J),J=1,N)
      WRITE(2,105) (Z(J),J=1,N)
      GO TO 9
   14 WRITE(2,120)
  120 FORMAT (1H1,15H CORE INTEGRALS)
      WRITE(2,103)ADIAG(1)
      DO 71 J=2,N
      JLOW=J-1
      WRITE(2,103)(A(J,I),I=1,JLOW),ADIAG(J)
   71 CONTINUE
      WRITE(2,121)
  121 FORMAT(/29H ELECTRON REPULSION INTEGRALS)
      DO 72 I=1,N
      WRITE (2,103)(A(J,I),J=1,I)
   72 CONTINUE
  100 FORMAT(5E16.8)
  101 FORMAT(7F11.8)
  103 FORMAT(11F7.3)
  104 FORMAT(I3)
  105 FORMAT(15F5.2)
    9 RETURN
      END
```

```
      SUBROUTINE SCOFH(N,F,FDIAG,GAM,Z)
      DIMENSION F(30,30),FDIAG(3),GAM(30,30),Z(30)
      WRITE(2,200)
200 FORMAT(/)
100 FORMAT(I2)
101 FORMAT(I1)
102 FORMAT(2I2,F7.3)
103 FORMAT(I2,F3.1)
104 FORMAT(I2,1X,F4.1)
  7 READ(7,101) LAB
      IF(LAB) 9,8,9
  9 IF (LAB-2)66,67,68
 66 WRITE (2,130)
130 FORMAT(16H F MODIFICATIONS)
      GO TO 69
 67 WRITE (2,131)
131 FORMAT (16H G MODIFICATIONS)
      GO TO 69
 68 WRITE (2,132)
132 FORMAT (16H Z MODIFICATIONS)
 69 READ(7,100) NITEM
      DO 20 K=1,NITEM
      GO TO (10,11,12),LAB
 10 READ(7,102)I,J,F(I,J)
      WRITE (2,102)I,J,F(I,J)
      IF(I-J)120,20,120
120 CONTINUE
      F(J,I)=F(I,J)
      GO TO 20
 11 READ(7,102)I,J,GAM(I,J)
      WRITE (2,102)I,J,GAM(I,J)
      IF(I-J)121,20,121
121 CONTINUE
      GAM(J,I)=GAM(I,J)
      GO TO 20
 12 READ (7,103)J,Z(J)
      WRITE(2,104)J,Z(J)
 20 CONTINUE
      GO TO 7
  8 DO 13 J=1,N
 13 FDIAG(J)=F(J,J)
      RETURN
      END


      SUBROUTINE SCOFI2(N,A,ADIAG,U)
      DIMENSION A(30,30),ADIAG(30),U(30,30),UTEST(30)
      DO 40 K=1,N
      ATEST=ADIAG(K)
      JTEST=K
      DO 41 J=K,N
      IF(ADIAG(J)-ATEST)42,41,41
 42 ATEST=ADIAG(J)
      JTEST=J
 41 CONTINUE
      ADIAG(JTEST)=ADIAG(K)
      ADIAG(K)=ATEST
      DO 40 I=1,N
      UTEST(I)=U(I,JTEST)
      U(I,JTEST)=U(I,K)
 40 U(I,K)=UTEST(I)
      RETURN
      END
```

```
      SUBROUTINE SCOFI3(N,M,F,FDIAG,T,G,Z,HAFGAM)
      DIMENSION F(30,30),FDIAG(30),T(30,30),G(30,30),Z(30),DIGR(30)
      DO 11 I=1,N
      DO 11 J=1,I
      RSUM=0
      DO 12 K=1,M
   12 RSUM=RSUM+T(I,K)*T(J,K)
      IF (J-I) 9,10,10
    9 F(I,J)=F(J,I)-RSUM*G(I,J)
      GO TO 11
   10 DIGR(I)=RSUM
   11 CONTINUE
      SUM=0
      DO 20 J=1,N
   20 SUM=SUM+DIGR(J)
      DO 21 J=1,N
   21 DIGR(J)=M*DIGR(J)/SUM
      DO 22 I=1,N
      RSUM=0.
      DO 23 J=1,N
      IF(I-J)24,23,24
   24 RSUM=RSUM+(2.0*DIGR(J)-Z(J))*G(I,J)
   23 CONTINUE
   22 FDIAG(I)=F(I,I)-HAFGAM+DIGR(I)*G(I,I)+RSUM
      RETURN
      END




      SUBROUTINE SCOFI4(N,F,FDIAG,T)
      DIMENSION F(30,30),FDIAG(30),T(30,30),FS(30,30)
      DO 8 I=1,N
      DO 8 J=1,N
      SUM=0
      DO 7 K=1,N
      IF(K-I) 5,4,6
    4 SUM=SUM+FDIAG(K)*T(K,J)
      GO TO 7
    5 SUM=SUM+F(I,K)*T(K,J)
      GO TO 7
    6 SUM=SUM+F(K,I)*T(K,J)
    7 CONTINUE
    8 FS(I,J)=SUM
      DO 10 I=1,N
      DO 10 J=1,I
      SUM=0
      DO 11 K=1,N
   11 SUM=SUM+T(K,I)*FS(K,J)
      IF(I-J) 12,13,12
   12 F(I,J)=SUM
      GOTO 10
   13 FDIAG(J)=SUM
   10 CONTINUE
      RETURN
      END
```

```
      SUBROUTINE SCOFK(N,M,F,T,G,Z,X,Y)
      DIMENSION X(96),Y(96)
      DIMENSION F(30,30),FDIAG(30),T(30,30),G(30,30),Z(30),R(30,30)
      DO 9 J=1,N
      DO 9 I=1,N
      SUM=0.
      DO 10 K=1,M
   10 SUM=SUM+T(I,K)*T(J,K)
    9 R(I,J)=SUM
      SUM=0
      DO 20 J=1,N
   20 SUM=SUM+R(J,J)
      DO 21 J=1,N
   21 R(J,J)=M*R(J,J)/SUM
      E=0.
      DO 8 I=1,N
      SIGK=0.
      DO 11 K=1,N
   11 SIGK=SIGK+2.0*(R(K,K)-Z(K))*G(I,K)
      DO 8 J=1,I
      BRAK=2.0*F(J,I)-R(I,J)*G(I,J)
      IF(I-J)6,7,6
    7 TERM=SIGK+BRAK+2.0*Z(I)*G(I,I)
      GO TO 8
    6 TERM=2.0*BRAK
    8 E=E+R(I,J)*TERM
      WRITE(2,100) E
      DO 1 J=1,N
      DO 1 I=1,J
    1 R(I,J)=2*R(I,J)
      SX=0
      SY=0
      DO 18 I=1,N
      P=R(I,I)-Z(I)
      SX=SX+P*X(I)
   18 SY=SY+P*Y(I)
      EMU=4.77*SQRT(SX*SX+SY*SY)
      SX=4.77*SX
      SY=4.77*SY
      WRITE(2,109) EMU,SX,SY
  109 FORMAT(/15H DIPOLE MOMENT=,F8.4,2X,5H XMU=,F8.4,2X,5H YMU=,F8.4)
      WRITE (2,102)
  102 FORMAT(/15H DENSITY MATRIX)
      DO 77 J=1,N
   77 WRITE(2,101)(R(I,J),I=1,J)
  100 FORMAT(/18H BINDING ENERGY E=,E16.8)
  101 FORMAT(10(F9.5,2X))
      RETURN
      END




      SUBROUTINE SCOFI5(MINK,LL,LH,N,XMIK,YMIK,C,X,Y)
      DIMENSION C(30,30),LL(64),LH(64),XMIK(64),YMIK(64)
      DIMENSION X(96),Y(96)
      DO 18 I=1,MINK
      JI=LL(I)
      JK=LH(I)
      SX=0
      SY=0
      DO 27 L=1,N
      C1=C(L,JI)*C(L,JK)
      SX=SX+C1*X(L)
   27 SY=SY+C1*Y(L)
      XMIK(I)=SX
   18 YMIK(I)=SY
      RETURN
      END
```

```
      SUBROUTINE SCOFL(N,MINK,AS,H,C,G,AT,I2,LL,LH)
      DIMENSION AS(30,30),H(30),C(30,30),G(30,30),Z(30),LL(64),LH(64),
     1AT(30,30),VECT(30)
200 FORMAT(8F10.5)
      IZ=1
      DO 8 MI=1,MINK
      DO 8 MJ=1,MI
      I=LL(MJ)
      K=LH(MJ)
      J=LL(MI)
      L=LH(MI)
      GM1=0.
      GM2=0.
      DO 44 IR =1,N
      C1=C(IR,I)*C(IR,J)
      C2=C(IR,I)*C(IR,K)
      DO 44 IT=1,N
      GM1=GM1+C1*C(IT,K)*C(IT,L)*G(IR,IT)
  44 GM2=GM2+C2*C(IT,J)*C(IT,L)*G(IR,IT)
      IZ=IZ+1
      IF(MI-MJ)26,27,26
  27 AT(MJ,MJ)=H(K)-H(I)-GM1
      VECT(MJ)=AT(MJ,MJ)+2.0*GM2
      GO TO 8
  26 AT(MI,MJ)=-GM1
      AS(MI,MJ)=-GM1+2.0*GM2
   8 CONTINUE
      DO 11 J=1,MINK
  11 H(J)=VECT(J)
      I2=0
      RETURN
      END




      SUBROUTINE SCOFP(MINK,LL,LH,H,I2)
      DIMENSION LL(64),LH(64),H(30)
      IF(I2-1)16,17,17
  16 WRITE(2,100)
 100 FORMAT(//34HSCF SINGLET CONFIGURATION ENERGIES)
      GO TO 18
  17 WRITE(2,101)
 101 FORMAT(//34HSCF TRIPLET CONFIGURATION ENERGIES)
  18 CONTINUE
      INZ=0
      INK=0
      K=1
      NMINK=MINK
1012 IF(MINK.LE.9) GO TO 1010
      INK=INK+9
      MINK=MINK-9
      GO TO 1011
1010 INK=INK+MINK
      INZ=1
1011 WRITE(2,1002) ((LL(M),LH(M)),M=K,INK)
1002 FORMAT(/9(1X,I2,2H -,I2,2X))
      WRITE(2,1009)(H(I),I=K,INK)
1009 FORMAT(9(F7.4,2X))
      K=K+9
      IF(INZ.EQ.0) GO TO 1012
      MINK=NMINK
      RETURN
      END
```

```
      SUBROUTINE SCOFM(MINK,AS,AT,H,C,I2,XMIK,YMIK,LL,LH)
      DIMENSION XMIK(64),YMIK(64),OSC(64),LL(64),LH(64)
      DIMENSION AS(30,30),H(30),C(30,30),AT(30,30)
      DIMENSION TRX(64),TRY(64)
      IF(I2-1)13,14,14
   13 WRITE (2,104)
  104 FORMAT(/15H SINGLET STATES)
      GO TO 15
   14 WRITE (2,105)
  105 FORMAT(/15H TRIPLET STATES)
   15 CONTINUE
      DO 9 J=1,MINK
      WRITE(2,1001)H(J)
 1001 FORMAT(/9HENERGY = ,E13.6)
      INZ=0
      INK=0
      K=1
      NMINK=MINK
 1012 IF(MINK.LE.9) GO TO 1010
      INK=INK+9
      MINK=MINK-9
      GO TO 1011
 1010 INK=INK+MINK
      INZ=1
 1011 WRITE(2,1002)((LL(M),LH(M)),M=K,INK)
 1002 FORMAT(/9(1X,I2,2H -,I2,2X))
      WRITE(2,1009)(C(I,J),I=K,INK)
 1009 FORMAT(9(F7.4,2X))
      K=K+9
      IF(INZ.EQ.0) GO TO 1012
      MINK=NMINK
    9 CONTINUE
      DO 11 J=1,MINK
      SX=0
      SY=0
      DO 10 I=1,MINK
      C1=C(I,J)
      SX=SX+C1*XMIK(I)
   10 SY=SY+C1*YMIK(I)
      TRX(J)=SX
      TRY(J)=SY
   11 OSC(J)=0.0000217*(SX*SX+SY*SY)*H(J)*8067.5
      WRITE(2,1003)
 1003 FORMAT(/20HOSCILLATOR STRENGTHS)
      WRITE(2,1004)
 1004 FORMAT(/5X,6HENERGY,5X,11HX-COMPONENT,5X,11HY-COMPONENT,5X,
     119HOSCILLATOR STRENGTH)
      DO 1005 J=1,MINK
      WRITE(2,1006) H(J),TRX(J),TRY(J),OSC(J)
 1006 FORMAT(/E13.6,3(F11.3,4X))
 1005 CONTINUE
      I2=I2+1
      IF(I2-1)6,6,7
    6 DO 18 I=2,MINK
      JUPP=I-1
      DO 17 J=1,JUPP
   17 AS(I,J)=AT(I,J)
   18 H(I)=AT(I,I)
      H(1)=AT(1,1)
    7 RETURN
```

7.5 REFERENCES

1. C. C. J. Roothaan, *Rev. Mod. Phys.*, **23**, 69 (1951).
2. E. U. Condon and G. H. Shortley, *The Theory of Atomic Spectra*, Cambridge University Press, 1935
3. J. C. Slater, *Quantum Theory of Molecules and Solids*, McGraw–Hill, New York, 1963, Vol. 1, p. 285
4. R. McWeeny, *Proc. Roy. Soc. (London)*, **A253**, 242 (1959).
5. J. A. Pople, *Proc. Phys. Soc. (London)*, **A68**, 81 (1955).
6. R. McWeeny and T. E. Peacock, *Proc. Phys. Soc. (London)*, **A70**, 41 (1957).
7. W. E. Moffit, *J. Chem. Phys.*, **22**, 320 (1954).
8. M. J. S. Dewar and L. Paolini, *Trans. Faraday Soc.*, **53**, 261 (1957).
9. J. E. Bloor, P. J. Daykin and P. Boltwood, *Can. J. Chem.* **42**, 121 (1964).
10. R. Pariser, *J. Chem. Phys.*, **24**, 250 (1956).
11. H. H. Greenwood and T. H. J. Hayward, *Mol. Phys.*, **3**, 495 (1960).
12. D. F. Evans, *Proc. Roy. Soc. (London)*, **A255**, 55 (1960).
13. J. R. Platt, *J. Mol. Spectroscopy*, **9**, 288 (1962).

Subject Index